Going Nowhere Fast

CRITICAL FRONTIERS OF THEORY, RESEARCH, AND POLICY IN INTERNATIONAL DEVELOPMENT STUDIES

Series Editors
Andrew Fischer, Uma Kothari, and Giles Mohan

Critical Frontiers of Theory, Research, and Policy in International Development Studies is the official book series of the Development Studies Association of the UK and Ireland (DSA). The series profiles research monographs that will shape the theory, practice, and teaching of international development for a new generation of scholars, students, and practitioners. The objective is to set high quality standards within the field of development studies to nurture and advance the field, as is the central mandate of the DSA. Critical scholarship is especially encouraged, within the spirit of development studies as an interdisciplinary and applied field, dealing centrally with local, national, and global processes of structural transformation, and associated political, social, and cultural change, as well as critical refections on achieving social justice. In particular, the series seeks to highlight analyses of historical development experiences as an important methodological and epistemological strength of the field of development studies.

ALSO IN THIS SERIES
The Aid Lab
Understanding Bangladesh's Unexpected Success
Naomi Hossain

Going Nowhere Fast

*Mobile Inequality in the Age
of Translocality*

SABINA LAWRENIUK AND LAURIE PARSONS

OXFORD
UNIVERSITY PRESS

OXFORD
UNIVERSITY PRESS

Great Clarendon Street, Oxford, OX2 6DP,
United Kingdom

Oxford University Press is a department of the University of Oxford.
It furthers the University's objective of excellence in research, scholarship,
and education by publishing worldwide. Oxford is a registered trade mark of
Oxford University Press in the UK and in certain other countries

© Sabina Lawreniuk and Laurie Parsons 2020

The moral rights of the authors have been asserted

First Edition published in 2020

Impression: 1

Published in the United States of America by Oxford University Press
198 Madison Avenue, New York, NY 10016, United States of America

British Library Cataloguing in Publication Data

Data available

Library of Congress Control Number: 2020937523

ISBN 978-0-19-885950-5

Printed and bound by
CPI Group (UK) Ltd, Croydon, CR0 4YY

*With thanks to everyone who helped along the way,
but above all to Ly Vouch Long*

Contents

List of Figures

1

Mobile Inequality in the Age of Translocality

If economic growth is the panacea of our age, then inequality is the zeitgeist. It is a part of everyday conversation and everybody, everywhere, has an opinion on it. What's more, these opinions often rank amongst people's most fundamental convictions, providing the conceptual foundations for entire world views. An aphorism such as Milton Friedman's famous contention that 'a society that puts equality before freedom will get neither. . . [whilst] . . . a society that puts freedom before equality will get a high degree of both' (Friedman and Freidman, 1980) continues to shape global discourse not as economic doctrine, but as a point of ideological conviction. Shorn of its quantitative pretentions, it adorns lists with names like '70 Great Quotations about the Glory of Honest Work and Achievement' in *Forbes* magazine (Powell, 2014), alongside Fredrick Hayek's (1960 [2014]: 77) assertion that 'the equality before the law which freedom requires leads to material inequality' and that foundational girder of Americana: 'early to bed and early to rise, makes a man healthy, wealthy and wise', set down by Benjamin Franklin in 1754 ([2007]: 85).

Yet, tendentious as it is, this is a discourse that is rarely presented as such. Whereas debates on poverty itself are more often framed in terms of fundamental rights and rooted in religion or rationalism—in other words, as positions of conviction—inequality retains an ostensibly technical foundation. Rather than being a question of justice, the economic dimension of its character cloaks it in a veneer of statistical objectivity, engendering a mysticism that undermines consensus. Debates about inequality are invariably undertaken quantitatively, rather than normatively and whilst it may appear intuitively appealing, this is an approach which shores up key defences of the libertarian argument, cloaking durable, visceral inequities in the anonymous language of growth.

In reality, inequality is an ideological battleground in which numbers are lent credence by rhetoric. This is an issue which has, in recent years, been described both as 'the biggest threat to the world' (Ghosh, 2013) and 'a blemish on Asia's growth story' (The Financial Times, 2014), yet its importance is continually dampened by a suite of opposing narratives. Discourse supporting GDP figures as the internationally recognizable hallmark of prudent stewardship remains largely dominant, bolstered by various narratives and axioms of development. Examples abound, but a popular perspective amongst economic liberals, for

Going Nowhere Fast: Mobile Inequality in the Age of Translocality. Sabina Lawreniuk and Laurie Parsons,
Oxford University Press (2020). © Sabina Lawreniuk and Laurie Parsons.
DOI: 10.1093/oso/9780198859505.001.0001

instance, holds that it is equality of opportunity, rather than income, which matters. Metaphorically speaking, 'a rising tide lifts all boats', even if transient waves of fortune leave some looking up or down at others for the time being.

Like many economic metaphors, this one carries considerable weight in shaping the direction of global change, but it is no truer for it. Beyond broad brush stories of national development, it is continuities and immobilities that stand out above all else. Yet debates over inequality versus income obscure this, revealing only 'chicken and egg' circularity, where the simple answer is both. Opportunity is an asset unequally shared, but also unequally realized. The best off have the resources to recognize and secure gains from the most fleeting of chances, whilst the poorest are left to grasp at even the surest bet, outcompeted as they are likely to be by their less well positioned, but better resourced peers.

This recognition is far from new, having attracted particular attention in the post-war European context of the welfare state. At that time, the British sociologist Paul Willis articulated the question of inequality in concise, almost tautological form: why, when educational reforms enacted in the 1950s and '60s had supposedly levelled the UK's social playing field, do working-class kids continue to get working-class jobs? As he and others (e.g. Levin, 1976; Neelsen, 1975) responded, the answer was simple: much of wealth cannot be measured in terms of physical, fungible, assets, but exists in the intangible form of narrative. The cultural tropes and shibboleths of one group—vocabulary, habits, interests, and hobbies—are praised, whilst those of another are denigrated, rendering discourse, narrative, and prejudice active components within the structures that protect wealth and inhibit its distribution. Stories and hard currency, otherwise put, are two peas in the pod of social difference.

The impact of such nuanced interpretations of inequality has been significant, with fine-grained analyses of inequality in this mould having long become key tools within the sociological and geographic literatures. Yet decades later they have made little impression on conventional inequality measures, which continue to be based on crude financial wealth metrics calculated predominantly at the national scale. When news outlets, NGOs and public figures proclaim that 'inequality is the biggest threat to the world' (Ghosh, 2013) or 'the greatest threat to democracy' (Obama cited in Bourke, 2016), therefore, they are making their assertions on the basis of an incomplete picture. In reality, the grip of structural inequality on people's lives is far more insidious than even proponents of its importance have thus far understood.

Indeed, the metaphor of being gripped, or held in place, by structural factors is key to conceptualizing inequality in an increasingly mobile world. Whether or not this is indeed the 'age of migration' as it has been called (Castles et al., 2013)— from a historical perspective the age of global, mass migration may be better attributed to the years between the mid-nineteenth century and the 1920s for instance (McKeown, 2004)—it is undeniably an age of unique mobility.

Technology has not only facilitated physical travel for more people, with greater ease and with lower cost, but conceptual movement and redistribution also. Driven by low (or sometimes no) cost international telecommunications, mind-sets, livelihoods, and households have taken on a diasporic character. Families and communities have had to adapt to retain meaning and relational coherence even as they have become detached from their spatial moorings. Intimate relationships are performed through global capital flows and international markets, so that for mothers living thousands of miles from their children, 'sending dollars shows feeling' (McKay, 2007: 175) and 'dutiful daughters' (Derks, 2008: 170) are made in urban factories, often hundreds of miles from their rural homes.

Complex though it is, the result of such dynamic interconnectedness amongst people and places has resulted in a translocal system which is more than the sum of its parts. Moving beyond the concept of transnationality, from which it emerged, translocality means multiple mobilities certainly, but it also means not 'losing sight of locality' (Oakes and Schein, 2006: 1). Translocality thus imagined describes a 'groundedness during movement' (Brickell and Datta, 2011: 4), as well as movement itself: localities mutually enhanced and shaped by their interlink-ages. For the purposes of this book, therefore, translocality is defined as the dynamic interconnectedness of localities linked by people, their actions, and activities, a framing that seeks to place human agency at its centre, whilst also accepting the role of non-human and structural factors in driving the systems within which mobility occurs. Appealing to this tethered fluidity of contemporary mobility, translocality as conceived herein is therefore a call to decentre bound-aries from the study of human movement, recognizing instead a world in which few barriers remain impervious to systematic flows of information and affection.

Nevertheless, amidst all this dynamism of mind and body alike, social mobility remains minimal for the vast majority. Between 1988 and 2011, the incomes of the poorest 10 per cent have risen by, on average, $3 per year, whilst members of the richest 1 per cent have seen their incomes rise 182 times as much (Oxfam, 2016). Moreover, these victims of economic stasis are neither unusual nor isolated. Not only have the incomes of the world's best and worst off raced apart to such an alarming extent that the planet's poorer half now possesses assets of equivalent value to its richest eight men (Oxfam, 2016), but these unlucky 3.7 billion souls are no strangers to the market and modernity. Rather, growing inequality viewed globally, as through any scalar lens, is a function of, rather than an exception to market forces. That the annual salary of a FTSE 100 CEO equates to that of 10,000 Bangaldeshi factory workers tells its own story. Entry to the modern sector means entry to a system that is itself vastly unequal.

This last example is especially pertinent. Labour migration has transformed the global economy in recent years, moving growing numbers of people into translo-cal status, as income generation becomes more and more spatially segregated from caregiving. Yet, 'surprisingly little research addresses directly the role of social inequalities' in the migration process (Faist, 2016: 323) and the 'rising tide of absolute global income inequality' (Goda and Torres García, 2017: 1051) has rarely been considered in this light, even as the two phenomena—translocal and migratory livelihoods on the one hand and growing inequality on the other—have moved increasingly to the centre of national discourse.

Consequently, much of the inequality within such systems remains unseen due to its persistently atomistic, static, and income-driven conceptualization. As Thomas Piketty (2015) recently explained to global acclaim, it is assets rather than income that underpin wealth, but even this great economic leap forward is insufficient. Measuring assets in terms of capitals—including such models that incorporate the various rough and ready definitions of social capital—elides the greater part of its character. Though relative, inequality is not an individually mediated state. Rather, it is intertwined and enmeshed in social relations in the form of norms, expectations, and morals; an ephemeral ballast to the fungibility of capital.

It is this very intangibility that allows inequality to be transmitted with such effectiveness between and within places and which allows it, in effect, to transcend mobility itself. Whether conceptualized as migration systems, chains, or flows, the movement of people cross-cuts and interacts with differences in wealth. These differences remain alive and active and are often entrenched rather than eroded by the mobility of their subjects. Moreover, such differences are nebulous only to the external observer. For many poorer migrants, translocal pressures are ever-present and all-consuming. Thousands of miles cannot dim the dread of 'the early morning phone call' from a family or friend in need (Lindley, 2010: 1), engendering differences in migrant livelihoods that are often expressed in leisure practices and patterns of association rather than incomes. Consequently, 'the rich migrate, just the same as the poor. But there is something different. When they go out [to the city] they stay separate and when they return they stay separate too' (Cambodian garment worker cited in Parsons, 2017).

Displacement, in other words, does not equate to solidarity, as has often been implicitly assumed in migration studies. Instead, migrants—voluntarily or otherwise—are subject to a myriad ongoing characteristics and differences that structure their interaction with each other and their host community. Being a garment worker, therefore, does not mean possessing an equivalent livelihood to one's colleagues, even those earning an identical salary. Rather, obligations are every bit as important as incomes in this respect. International and domestic financial commitments mean that she with the failing farm and the ailing father is incomparably worse off than she whose relatively comfortable household allows her simply to retain, spend, and save her own salary for herself and her nuclear family.

Moreover, these differences are expressed in far starker terms than money. Inequality is not abstract and numerical, but 'fleshy' (Dixon, 2014) 'embodied'

(Casanova, 2013: 561) and lived. Through the scars of exploitative labour and the corporeal demarcation enacted by hunger, inequality is 'written on the body' (Baviskar, 2001: 354) of women and men. Even the unborn are similarly marked, as translocal patterns of differentiation affect maternal health and patterns of pre- and post-natal care in the short term (WHO, 2015; Daoud et al., 2014) and educational and life outcomes over a longer timeframe (Härkönen et al., 2012).

Viewed thus, inequality is something fundamental, cloying and penetrative, belying teleological notions of growth, progress and—perhaps especially— migration. But it is also fluid and adaptable in form. Capability is structured by perception and what it means to be worse off by definition differs hugely across contexts, yet this inherently relative, cultural, and local phenomenon is able to persist in the widely spaced and mobile context of contemporary mobile liveli- hoods. Disadvantage changes its form to suit its environment, but in doing so retains its structure and relative hierarchies. The poor in one place therefore invariably become the poor in another, but key questions nevertheless remain. How have translocal patterns of living served to stratify and distinguish people working in similar ways from each other? What is the catalyst driving translocal livelihoods apart? Why, above all, does mobility engender stasis in some, but not others?

The answer, we argue, is narrative. Stories, prejudices, and 'imagined commu- nities' (Anderson, 1983: 207) set the boundaries of the possible, determining who is in and who is out, who deserves and who doesn't and which behaviours are desirable and which aren't. Through overlapping intersections of discursive exclu- sion, the poor are isolated, subdivided, and justified. Above all, their (relative) poverty is made to seem natural and it is this very narrative permanence that allows it to persist under rapidly changing and dynamic conditions. Denigration, we find, is no respecter of context and the tropes of inequality that distinguish one group from another are soon re-established and entrenched across translocal boundaries. These are rarely expressed in economic form, but more often as aesthetic or moral narratives which include scatterings of praise to engender the illusion of balance.

The classificatory distinction between 'expatriates', who migrate from wealthy countries, and 'immigrants', who tend to have come from poorer ones, is a famous example of this phenomenon, but such distinctions do not begin and end with the nation state. Rather, in determining the warmth (or otherwise) of a migrant's welcome, wealth differences at origin matter within as well as between countries, with 'a rather complex structure' (Tsuda, 1998: 317) of social, ethnic, and eco- nomic factors contributing to the form and extent of discrimination. From the positioning of wealthy and successful Indians as 'symbols of a re-worked "model minority" myth . . . at precisely the moment when poorer immigrants were being targeted as unworthy of membership in America' (Shukla, 2013: 176), to former British Prime Minister David Cameron's description of asylum seekers entering from Calais as a 'swarm' in 2015 (Dahliwal and Forkert, 2015: 50), classifications of 'right' and 'wrong' kinds of immigrants are based on assumptions people share about 'who contributes to the welfare of the nation and who does not' (Newton, 2005: 139).

However, that denigration is both a cause and a consequence of poverty is not a new observation. Not only have ideas of the 'deserving poor' been discussed for hundreds of years, from twelfth-century manuscripts to the present day (Erskine and Mackintosh, 1999), but systems of identification have long accompanied these. In the Victorian era, for example, a system of badging purported to differentiate those whose poverty was genuine—meaning outside of their control—from those whose poverty was viewed as personally induced and hence natural (Hindle, 2004). In historical context, the exclusion of a sizeable group of people in this way was no anomaly, but merely formalized and extended a persistent trend whereby discourse is utilized to delineate the 'universe of obligation', as Helen Fein (1979) puts it, outside of—or at the margins of which—may be deemed to fall a certain fraction of the community.

What is novel in the current era is the multi-locality with which such narratives emerge and persist. Families and individuals are no longer atomistically knowable; an assessment of their assets requires information from multiple sites. Yet narratives of denigration persist between places and people. Migrants bring with them not only assets, but customs, attitudes, and hierarchies also, which are not wiped clean by travel and context. Rather, they become channels through which to direct broader narratives of exclusion, leaving the migrant members of poorer families as the primary recipients of denigration aimed purportedly at migration in general.

The Western world provides ample example of this phenomenon. Nodal debates concerning the refugees from Syria and the potential rectitude and efficacy of Donald Trump's proposed border wall between the United States and Mexico have catalysed long running debates on citizenship, belonging, and the rights of non-citizens. What it means to be a migrant, and what a migrant should be, are questions that have been at the centre of a global discourse that continues to view human mobility as an exception to the static norm required of nation states. Nevertheless, the global nature of these phenomena renders detailed analysis problematic. From the large-scale vantage point necessary to appreciate the scale of these migration systems, as well as the genus of the narratives that underpin them, much of the nuance of their operation is missed.

What escapes attention in particular is that the structural and narrative components of migration systems are not discrete, but closely linked and mutually constitutive. Prejudice both at home and destination is predicated not on 'traditional' or 'conservative' views in isolation—although it may draw on and be informed by the lexica associated with those terms—but by systematic factors engendered in multiple locations. Both negative and positive attitudes to migrants therefore have their roots in historical patterns of mobility and the (local and global) economic conditions that engender them, as well as specific contemporary conditions. Economics, mobilities, and discourse therefore exist in constant, triangular interaction, ensuring that disadvantage in one place and of one type is replicated and enhanced in others.

Nevertheless, so many and various are the structural dynamics of global migration systems that subtle redirections such as these can be challenging to perceive. As various authors have noted, the largest geopolitical processes manifest also at the smallest scales (Brickell, 2014; Pain, 2014; Smith, 2012), making a fine grained appreciation of culture, gender and community dynamics essential tools in mapping the multi-scalar and cross-contextual impact of mobility. Yet far larger-scale constraints on migration have, of late, become the norm, garnering such substantial and sustained attention as to dominate the narrative of human movement (Rigg, 2013; Elmhirst, 2012; King and Skeldon, 2010). Aiming to redress this balance, this book eschews the international viewpoint, taking instead the small, economically integrated and highly mobile country of Cambodia as a microcosm of a worldwide phenomenon. The aim from this standpoint is to draw out the cohesive narrative of Cambodian migration from a diverse body of evidence, bringing together diverse thematic foci to highlight the changing meaning of mobility as it straddles the rural–urban continuum.

Contextually, Cambodia has a particular advantage in this respect. Its relatively homogenous ethnicity and culture—in comparison to the region, though somewhat exaggerated in national and international discourse (Ter Horst, 2008; Edwards, 2007)—and tightly wound migratory linkages render the phenomenon visible with unusual clarity. Indeed, migration has in recent years become so fundamental, contentious, and transformative a phenomenon here that the influence of mobility is today felt in every sphere of life and livelihoods. From the shifting role of women and girls in a rural society increasingly dependent on modern sector remittances (Parsons et al., 2014), to the marketization of grandmothers' care-giving roles (Parsons and Lawreniuk, 2017) to the rise of novel dating practices amongst young migrants and all the narratives of love and denigration that go with them (Parsons and Lawreniuk, 2017), Cambodian society is imbued from top to bottom with the meanings and structures of mobility.

Furthermore, the detailed focus allowed by the relatively small scale of Cambodian migration systems permits unusual insight into the stratification of these shifting practices. Neither economic, cultural, nor indeed ecological change is experienced in a direct manner here, but rather through the lens of assets and endowments which link these three areas together. Thus, whilst the best off migrants build networks and progress in the urban space, their poorer colleagues are left buffeted and immobilized by rural and urban pressures. Agro-ecological shocks to the family farm force remittances—already averaging over 50 per cent of gross salary amongst this group—yet further up, leaving workers to survive seventy-two hour weeks on mean portions of the cheapest food available.

Chronic illness for those thus afflicted is both rife and compounded by an isolation borne of their inability to pay for the smallest of courtesies (Parsons, 2016). A mere half a hectare of land, or a few hundred metres distance from a water source, can therefore make the difference between urban advancement and being trapped in the indefinite limbo of factory work to service debt repayments; indentured labour by proxy.

Unlike inequality more generally, therefore—a term so broad as to be viewed predominantly in relation to its opposite: a lack of equality, or the absence of sameness—translocal inequality may be defined in its simplest form as that process whereby wealth discrepancies in one place engender differences in livelihoods elsewhere. More specifically, it constitutes a reframing of the subject of inequality away from the atomistic individual—or similarly the individual within a locally bounded household—towards communities grounded in mutual fields of obligation. Viewed thus, translocal inequality means an unequal distribution of resources and opportunities which cuts across local boundaries, and is structured and sustained by the actions of multiple people in multiple places simultaneously.

Above all, though, the key implication of adopting a translocal lens on inequality is a recognition of its embeddedness in the networks that interconnect place. To view inequality translocally is to recognize the complex interconnection of social, cultural, and economic factors and thus to see it situated inherently in multiple, linked, spaces, rather than moving from one to another. Unlike economic inequality, therefore, which denotes unevenness of income or assets between individuals or groups; or social inequality, which indicates the unequal access to opportunities and rewards engendered by the balance of stigma and status (Marger, 1999), translocal inequality means the interaction of these forms of inequality across space and between co-constitutive places.

Nevertheless, the aim here is not merely to demonstrate the existence of this phenomenon, although ample evidence will be provided in this respect. Rather, the intention is to highlight how changes in one aspect of life and livelihoods may—through the medium of people, their obligations, stories, and advantages—manifest in quite different spheres. Translocal inequality is therefore by definition also fungible inequality: disadvantage manifest in and portable to multiple spheres of life, both economic and non-economic.

Viewed thus, that the loss of a family rice crop to drought in the countryside should lead to accusations against a migrant worker of being a 'rice person' in the city is neither coincidental nor unusual: controlling assets means controlling narrative and vice versa. Only the better off acquire the knowledge and tools to flaunt the traps of denigration associated with belonging and exclusion, leaving others to build mobile livelihoods on a tenacious bedrock of prejudice. Translocal inequality is therefore total inequality. Rendered natural in appearance by narrative, it imprints itself indelibly on mind as well as body, structuring future outcomes, lives, and livelihoods with minimal regard for locality.

Drawing on a decade of empirical research, the goal of this book is therefore to elucidate the complexity and multiplicity of human differentiation. In contrast to the cross-sectional approach favoured in the most influential recent literature on inequality (Piketty, 2014 and Dorling, 2014 being major examples), therefore, this book employs a methodology concerned with linking the multiple intersections of inequality within a single national context in the global South. It draws on fieldwork conducted in the course of five linked, but empirically distinct research themes exploring dimensions of inequality in Cambodia and ultimately constructing a wider picture from the bottom-up aggregation of insights. In each case, the methods employed—and detailed in greater depth in Chapters 3 to 7—employ a combination of qualitative and quantitative methods to evidence how each facet of inequality operates in a linked and dynamic manner between spaces.

Although the themes addressed in each chapter employ distinct combinations of methods, the guiding methodological principle has been the use of linked multi-site data collection: an approach intended to refocus and mobilize the study of inequality. In Chapter 3, for example, this takes the form of a rural socio-economic survey, into which method was integrated follow-up interviews with 49 out of the 50 migrant workers originating from surveyed households. In Chapter 4, by contrast, the findings of four linked multi-site studies are combined in search of integrated insights into the unequal power relations lived by women in translocal households. Chapter 5, similarly, brings together insights from three studies to produce a multi-sited ethnography of resource use across translocal systems, whilst Chapters 6 and 7 draw upon smaller-scale and more intensive approaches, focusing on how stigma and prejudice shape the mobilities of marginalized groups across rural and urban areas.

As such, the research employed in this book is the result of an iterative mixed methods (De Lisle et al., 2017) approach, whereby methodological components are integrated in order to answer key thematic questions—i.e. how to rural assets shape the experience and practice of urban work—and subsequent studies seek to build on and develop the findings of these questions. The aim here is therefore to apply a 'range of methodological tools which can capture not just the economic exchanges, political organizations or social networks across sites of departure and destination, but also the negotiation of wider range of spaces and place in between' (Brickell and Datta, 2011: 4). By grounding its conception of inequality within these interlinked localities, a conception of inequality is sought which is fully embedded in the context of unequal dynamism.

Nevertheless, in order to view inequality as the 'total social fact' (Mauss, 2002 [1925]: 100) we argue that it constitutes, it is necessary, first, to highlight how existing conceptions have fallen short. Chapter 2 of this book will therefore undertake an examination of conventional measures of inequality, in order to highlight in broad terms what is missed by the large-scale economic viewpoint that is almost ubiquitous to its study. Therein, we will begin by critiquing dominant inequality measures such as the Gini, Palma, and Thiel indices, before highlighting key areas—such as health, education, and the environment—in which the shortcomings of these approaches are especially pronounced. As such, this chapter will lay both the conceptual and empirical groundwork for what follows by outlining the space necessary for a new perspective on inequality.

Having done so, Chapter 3 will outline the central theoretical position of the book. It uses data from a rural village in Kandal province and its diaspora of migrants to Phnom Penh and other peri-urban areas to demonstrate the extent and importance of the social linkages between sender and destination areas. In doing so, it aims not only to demonstrate that inequality is transmitted from one place to another, but that it is co-created in both places simultaneously. Social relations are key to this position, not only as a means to coordinate resource distribution across multiple locations but also as a resource in themselves. Simply put, the economic freedom afforded to the better off generates social returns which greatly enhance the future prospects of migrants and their households. This differential underpins not only inequality of mobility, but—more fundamentally—translocal inequality. As this chapter will show, migrants' liveli-hoods are determined by their households' specific endowments and needs as directly as if they remained within earshot; the worst off existing in state of perpetual isolation and penury, whilst the better off enjoy the freedom of the city in their leisure time, building friendships, spending and investing in a manner directly influential on their future. Most notably, these distinctions are not binary, but finely graded, with each notch of family wealth bringing with it a different set of circumstances, sacrifices and future opportunities.

If these first three chapters outline what is meant by translocal inequality, Chapter 4 begins to address the mechanisms by which it operates. Retaining the Kandal migration system outlined in the previous chapter, it begins to extend the concept of translocal inequality from the realm of economy to that of ecology. Thus, it shows how the patterns of mobile disadvantage evidenced in the previous chapter manifest in changes to cropping patterns and resource use. However, in keeping with the overall framework of the book, it views these through the lens of the social dimensions at play. Changing livelihoods means changing patterns of labour distribution and these in turn drive change within the household. Who wins, who loses, and how, will be the undergirding theme addressed in a chapter which focuses on the changing characteristics of gender in a marketizing and increasingly translocal agricultural context, as well as how new norms and modal-ities of ageing, which have arisen in response to the pastoral needs of a translocal household. In both cases, it will be shown, mobility has become integral to narratives of duty and moral worth, leaving those excluded from the translocal agro-economy often in a position of active immorality.

These themes are developed and extended in Chapter 5, which broadens the lens used in the previous chapter to explore Cambodia's translocal ecologies at the national scale. From this perspective, it will first examine the direct ecological impact of labour migration on receiving communities through the testimonies of migrants and 'indigenous' residents of a peri-urban garment enclave to the South of Phnom Penh. By applying a translocal perspective to land use change, these accounts will be used to consider how the manifestation of economic forces through mobility generates both short and long term impacts on the environment. Moreover, this relationship is considered bi-directionally, incorporating the impact of remittances on sender ecologies in a linked and dynamic manner. On this basis, the chapter will conclude by considering the relationship between inequality and ecological change. Although no two people experience climate shocks in the same manner, it will argue, inequality is both a consequence and a key driver of ecological degradation, encouraging changes in land use and liveli-hoods practices that in many cases cause harm to the local and translocal environment.

Having established the mutually constitutive role of ecology and mobility, Chapter 6 presents a further layer to the analytical structure thus far established. Introducing a thematic dyad that continues in Chapter 7, it begins to explore how the translocal systems of mobility explored thus far contribute to the production, revision and use of cultural narratives of inclusion and exclusion. The chapter focuses on begging migration between Prey Veng and Phnom Penh, centring on a resurgent folk tale of a cursed village whose inhabitants—rich or poor—are forced by evil spirits to go begging for alms each year. In moving beyond the supernatural dimensions of this narrative, it uses accounts from both Phnom Penh and the sender villages most associated with the myth to show that this story and others like it should not be viewed not as entities isolated by the boundaries of culture, but as integrated components within wider translocal systems. As such, narratives such as these are not merely symptomatic of translocal structures, but actively involved in driving them and determining their form. Discourse therefore exists in dynamic conversation with the structural components of mobility, rendering narrative a powerful factor in the genus and persistence of exclusion and inequality.

Building on this discourse driven mobility framework, Chapter 7 will once again leap scales to explore the wider implications of this lens. In doing so, it interrogates the linkages between Cambodia's resurgent political opposition and the Kingdom's large-scale union movement, using quantitative and qualitative data to demonstrate how a recent resurgence of nationalism within political discourse is linked to the changing mobility of the electorate. Indeed, whilst the highly mobile members of the labour movement are increasingly prominent figures in national politics, the political discourse of their support engenders immobility and exclusion from translocal systems, directed in particular against those with Vietnamese ethnicity. The targets of political prejudice are therefore determined in large part by perceptions of their translocal behaviour; of their having moved in an incomplete or immoral manner which undermines their claim to a place in the Cambodian nation.

In highlighting how national identity and the right to belong are bound increasingly to mobility, whilst immobility increasingly means exclusion—formal and informal—from the moral universe of the nation, this final empirical chapter begins to draw together some of the strands outlined in the preceding pages, leaving the conclusion, Chapter 8, to contemplate the viability and value of a translocal inequality framework. Therein, the three lenses though which the translocal dimensions of inequality have been explored—economy, ecology, and discourse—are considered in concert. The linkages between each are revisited and the possibility of a cohesively translocal conception of inequality explored. After considering the theoretical implications of integrating discursive and more systematic frameworks of migration, the chapter concludes by highlighting how the holistic approach employed here may be extended to produce further insights, as well as suggesting key areas which could potentially benefit most from such a lens.

As such, by demonstrating the durability of wealth structures in highly mobile conditions, this book presents a case for greater attention to the small scale, often intangible, manifestations of inequality. However, this is a case made through example rather than abstraction. Narratives are inherently contextual and Cambodia's socio-cultural environment is therefore central to this analysis, as it must be to any effective account of translocal inequality. From garment workers to fishers and farmers, Cambodians live within webs of mutual dependency that structure their well-being both now and in the future. What the venerable commentator Meas Nee once called this 'net of obligations' renders opportunity and income two sides of the same coin. Yet a failure to fully grasp this has often undermined accounts of deprivation in the country. Development here—as in much of translocal Asia and beyond—is not the aggregation of individual experiences, but a contest played out between groups, families and networks. Only by recognizing the importance of these linkages does the true nature of inequality emerge. No rising tide can lift a boat whose 'net' has become a chain.

2

The Fallacy of Macroeconomic Indicators

During the past half century, Cambodians have witnessed continuity in almost unimaginable change. Liberation from the nightmare of Democratic Kampuchea was followed by ten years under Vietnamese dominance as the People's Republic of Kampuchea [PRK]. When control was subsequently passed to the UN Transitional Authority in Cambodia to facilitate elections in 1993, a brief period of political uncertainty preceded a bloody coup in July 1997. Thereafter, the beneficiaries of that power grab, Prime Minister Hun Sen and his party the CPP spent almost twenty years gradually consolidating their position, as political opposition withered and fell away. The Cambodia National Rescue Party's [CNRP] formation briefly reversed this trend, garnering 44 per cent of the popular vote in the 2013 election and 2017 local elections. Yet the return of democratic pluralism proved short lived. The CNRP's dissolution in 2017 marks only the latest twist in a tale that has rarely veered far from autocracy.

Nevertheless, this near half century of political upheaval has included a substantial period of economic recovery. Cambodia's growth since its first post-conflict elections in 1993 has been described as a development 'miracle' (Madhur, 2013: 1), as the country has achieved 'much more rapid economic development over the past two decades than even the most optimistic forecasts could have projected at the time of the 1991 Paris peace settlement' (Hill and Menon, 2013: 64). Gross domestic product more than sextupled between 1993 and 2018 (World Bank, 2018), including a nine-year period between 2004 and 2012, during which 'growth was tremendous, ranking amid the best in the world' (World Bank, 2014: x). At the same time, poverty and development indicators have shown substantial improvement, with poverty more than halving, from 53.2 per cent to 20.5 per cent (World Bank, 2014). These figures have combined to produce a remarkable quarter century, in which:

> Between 1990 and 2015, Cambodia's HDI value increased from 0.357 to 0.563, an increase of 57.7 percent ... Between 1990 and 2015, Cambodia's life expectancy at birth increased by 15.2 years, mean years of schooling increased by 2.0 years and expected years of schooling increased by 4.2 years. Cambodia's GNI per capita increased by about 277.9 percent between 1990 and 2015.   (UNDP, 2018: 2)

The World Bank holds these statistics to indicate 'pro-poor' growth throughout this period (World Bank, 2014). However, the extent to which it has been

Going Nowhere Fast: Mobile Inequality in the Age of Translocality. Sabina Lawreniuk and Laurie Parsons,
Oxford University Press (2020). © Sabina Lawreniuk and Laurie Parsons.
DOI: 10.1093/oso/9780198859505.001.0001

equitable is 'less clear' (Hill and Menon, 2013: 57). Bearing in mind the unusually equal distribution of resources which existed at modern Cambodia's inception—brought about by a stagnant economy and the egalitarian redistribution of arable land amongst the population in 1989 (Yagura, 2015)—the Kingdom's subsequent development might be viewed as anything but progressive. Indeed, 'by some accounts, it was one of the most rapid increases in inequality in developing Asia' (Hill and Menon, 2013: 57–8).

For those who take this view, inequality is often ascribed to weak domestic institutions, which have allowed 'a kleptocratic elite' (Global Witness, 2007: 6) alarming control of the nation's resources. Patronage and patriarchy have under-pinned a systematic 'redistribution of resources from the poor to the rich' [on which] 'Cambodian tycoons linked to the ruling party are well placed to cash in' (Hughes, 2008: 72), thereby strengthening the hold of the 'almost almighty' (Deth and Bultman, 2016: 87) prime minister still further. As a result, 'increasing landlessness, or the transformation of peasants into "living ghosts", is a threat that many Cambodians face' (Schneider, 2011: 27).

There is more than a little truth to this perspective. Certainly, institutionalized corruption has wrought substantial harm to the equity of Cambodia's private and (nominally) public resources alike. Nevertheless, to place local institutions at the root of this process of redistribution would be misrepresentative. In a 'highly open economy' (Hill and Menon, 2013: 49) such as Cambodia's—subject to regional and international economic frameworks at a number of scales—no phenomenon exists in isolation. Inequality in Cambodia is as much a product of geopolitical (Brickell, 2014) and ecological (Ek, 2013; Ros et al., 2011) processes and structures as it is of domestic ones. Yet 'the structural inequalities of capital are increasingly misrecognized' (Springer, 2010: 931) as idiosyncratic, rather than systematic, features of the contemporary Cambodian economy; the product of corruption, nepotism, and transitional development, rather than wider geopolitical forces.

In large part, this derives from the use of economic conceptualizations that cross-cut and abstract key axes of inequality in the Kingdom. Cambodian inequal-ity is predominantly discussed in four ways: in relation to macroeconomic indi-cators, health, education, and land, reflecting its conceptualization according to the economic triumvirate of economic, physical and human capital. However, the continued reliance of such models on 'dual sector' conceptions of migration and development (Harris and Todaro, 1970; Todaro, 1969; Lewis, 1954) generates problematic analytical flaws. Like poverty, inequality is multi-dimensional (Li and Wei, 2014; Justino et al., 2004) and mobile (Parsons, 2016; Jensen and Richardson, 2007; Olvera et al., 2004). It exists 'along urban–rural lines' (Beresford et al., 2004: 12), but also between them, as advantages in one sphere manifest fungibly in others.

For contemporary Cambodians, therefore, the manifestation of rural and urban inequality are not discrete phenomena, but two elements within a holistic and multi-local structure of wealth. Bi-directional flows of people and money link the livelihoods of rural and urban areas, ensuring that ecological stresses and urban prejudices alike are felt in both spheres as they manifest. Moreover, these pressures are not only economic, but also embodied. Rising urban rents and agricultural shocks combine to impact health and education in a manner that inter-temporally and inter-generationally entrenches the idiosyncratic vulnerabilities of the worse off, structuring them in a manner that reflects and overlaps with existing social systems.

Indeed, the social structuring of inequality is a crucial feature of its manifestation. Narratives of value, good conduct, desirable actions and appropriate interactions underpin economic differences in a manner that eschews externally imposed categorizations. Gender norms of 'dutiful daughters' (Derks, 2008: 170) working in the factories to support their families, whilst profligate sons squander their salaries, structure spending patterns on a household basis; beggars' mobility and livelihoods are driven by stigma at home and in the city (Parsons and Lawreniuk, 2016a); and stories of nostalgia, both domestic and international, bind paratransit workers into decades long cyclical migrations whilst all around them motorized transport overwhelms their livelihoods (Parsons and Lawreniuk, 2016b).These hierarchies and patterns of inclusion and exclusion render intelligible what would otherwise be too complex and diverse a milieu. They are the patterns of shared thought around which goods of every type coalesce.

From this standpoint, this chapter will proceed to explore this complex landscape, outlining the evidence for inequality in its numerous formulations. It will do so according to the broad outline of conventional economic conceptions, examining economic, physical, and human capital in turn. Thus, it will outline first the shortcomings of conventional means of measuring inequality before turning, secondly, to a translocal analysis of land in its relations to wider, multi-faceted development processes. Thereafter, the final part of the chapter will highlight inequities in health and education, exploring the interconnected manner in which these embodied inequalities manifest in—and are sustained by—broad geopolitical processes.

2.1 Traditional Measures of Inequality

Cambodia's growth in recent years has been the source of much congratulation, as international organizations acclaim the Kingdom's 'hard won macroeconomic stability and steady adherence to prudent policies', which have allowed it to 'become one of the world's fastest growing economies in the first decade of this century' (Unteroberdoerster, 2014: 1). Moreover, this period of growth is frequently argued to have benefited the country's poorest citizens, with numerous commentators asserting that 'most of this growth benefitted the poor and as a result poverty rates dropped to 20.5 percent' (Buehler et al., 2016).

More specific indicators are also invoked. The UNDP (2016: 3) points to 'Cambodia's progress in each of the HDI indicators', noting substantial improvements in life expectancy, mean years of schooling, expected years of schooling and GNI per capita—amongst other indicators—as evidence of the benefits Cambodia's economic development has brought to those who most need it. In addition, Cambodia has outperformed comparable developing countries such as Myanmar and Papua New Guinea during this period, generating greater improvements in key indicators than its peers (UNDP, 2016).

Nevertheless, despite an impressive array of achievements, measuring the scale and impact of inequality is more difficult. Whilst the UNDP takes income inequality into account in certain of its measurements—stating for instance that income adjusted Human Development indicators [HDI] suggest an HDI loss of only 22.5% due to income inequality, a figure lower than the average for medium HDI countries (25.7%) and comparable states such as Laos PDR (27.1%)' (UNDP, 2016: 3)—these data import the generalizations of widely used economic measures. In particular, they are dependent on broad segmental comparisons of the population, an approach with well noted analytical shortcomings.

Most commonly, the metric used is the Gini coefficient, which ranks countries on a scale between 0 and 1 according to their deviation, on a Lorenz diagram, from a line of 'perfect equality' in which all resources are equally distributed. According to this measure, Cambodian inequality has exhibited a rapid and somewhat unlikely instability, with a sharp rise between 2004 and 2007 interrupting a generally equalising trend before and afterwards (World Bank, 2014; ADB, 2014). Moreover, alternative measures produce comparable results. During the period 1994–2009, the Palma Index, for instance—which produces a simple ratio of the top 10 per cent to the bottom 40 per cent of the population—declined from 1.7 to 1.5. (ADB, 2014: 10)

Similarly, Lorenz curves (on which Gini coefficients are based) and the Thiel Index, which is primarily used to highlight intra-regional and inter-regional attributions, reached comparable conclusions (JICA, 2010: 9).

As the ADB themselves admit, however, 'these figures must be interpreted with caution' (ADB, 2014: 9) due to the low likelihood that the Cambodia Socioeconomic Survey [CSES] on which they are based has been able to capture the upper reaches of Cambodia's income distribution. Indeed, 'despite a larger and more representative sample than used previously, the 2009 CSES measured average consumption in the richest quintile at just $3.75 per person per day' (ADB, 2014: 9). The country's richest people, in other words, are absent from these calculations.

More broadly, moreover, pressing questions have been raised (ADB, 2014; Yu and Fan, 2011; Sothearith and Sothearith, 2009) about how inequality has apparently continued to decline in the face of sharply rising rice prices during the period. Rice provides some 75–80 per cent of calories to Cambodians on average (UNDP, 2013) and 41 per cent of Cambodians are net buyers of rice (World Bank, 2016: 8), including most of the poorest strata of the population (ADB, 2014). Substantial rises in the cost of the staple, which remains considerably more expensive than that exported by neighbouring countries (IFC, 2015), may there-fore have been expected to put considerable pressure on rural livelihoods. However, analysts point to 'three countervailing factors' (ADB, 2014: 11) to explain the purported lack of impact:

> First, the returns to labor increased significantly: daily wages in rural areas more than doubled in 2007–2012, with a particularly sharp increase (57%) in 2007–2008. This helped many workers partly offset the impact of high food prices. Additionally, men and women workers alike benefited from increased wages. Second, rice yields also increased substantially. The average yield for wet season rice increased from 1.7 tons of paddy per hectare in 2004 to 2.9 tons by 2011. Third, income from off -farm self-employment also increased.
>
> (ADB, 2014: 11)

There is undoubtedly a good deal of truth to these explanations. Rural wages almost tripled in the decade from 2005 and 2015 (IBRD and World Bank, 2015), as urban alternatives began to tempt rural Cambodians away from agricultural wage labour. Closely linked to this process, wages and remittances from the modern sector have proved highly influential to rural economies. Data from the garment sector alone suggest that, across the industry as a whole, remittances account for almost 40 per cent of gross salaries, or more than $425,000,000 annually being sent by workers to their relatives, most of whom remain in rural areas (CARE, 2017).

Nevertheless, despite their use to explain counter-intuitive trends in nationwide equality data post hoc, these translocal conceptions of income and livelihoods are rarely incorporated into the process of collection or analysis. Geographical and spatial distinctions—in particular rural/urban, but also other administrative and agro-ecological categorizations—are used frequently as a means to analyse differ-ences in poverty metrics, meaning that 'the complexity of the poverty issue is too often hidden behind simplistic indicators and development goals' (Varis, 2008: 225). Consequently, it is often asserted that 'the poverty gap and severity in Phnom Penh are negligible' (JICA, 2010: 4), or that 'the largest component of overall inequality is inequality within rural areas' and 'between regions' (JICA, 2010: 9).

Spatially mediated conclusions such as this fail to recognize that poverty and inequality depend to a large extent on the people to whom one is connected and thus how income generated in one place may impact on livelihoods experienced in another. In a country where a third of the populations were estimated to be migrants as long ago as 2010 (NIS, 2010), since which time the number of employees in the garment industry has more than doubled (UNDP, 2016) alongside substantial rises in other migrant industries, such geographic distinctions are somewhat arbitrary. Consequently, accounts holding that 'the benefits of economic growth have been disproportionately urban, aiding the rich at the expense of the poor' (Hughes, 2008: 70) must be significantly nuanced. Urban growth (see Figure 2.1) may engender rural inequality, or vice versa.

Indeed, in high migration areas, urban growth is to a significant extent also rural growth: more jobs and rising incomes following the minimum wage campaigns of recent years have seen substantial flows economic flows directed towards these regions. At the same time, though, rural costs are also urban costs. Rising urban prices and rural pressures have increasingly squeezed consumption for both remitters and receivers of remittances (Lawreniuk, 2017; Parsons, 2017), meaning that droughts and floods are felt by factory workers as well as farmers and rising urban rents felt by rural villagers as well as those who pay them directly (Parsons, 2016).

As a result, both modern sector and rural work are defined by their mutual dependency. Factory workers' livelihoods, though relatively homogeneous in income terms, are vastly differentiated by the obligations of each rural household. Agriculture lives and dies by the remittances of migrant workers, as inputs rise in price and costs are met in advance through debt. Consequently, the money migrants send home is therefore often the only thing standing between a family and the loss of their land and factory workers' attitudes to their salary reflect this. Demonstrations and strikes become more frequent and voluminous amidst the pressures of peak agricultural periods, as when workers protest, they are doing so not for themselves, but for their families, home, and the preservation of their roots in the countryside.

This translocal character of Cambodian livelihoods has been highlighted by a number of authors (e.g. Parsons, 2016; Parsons and Lawreniuk, 2016; Bylander,

Figure 2.1 The rapidly expanding skyline of Phnom Penh, 2018.
Source: Courtesy of Thomas Cristofoletti/Ruom/Blood Bricks/Royal Holloway.

2015; Brickell, 2011), but its implications for inequality have yet to be fully incorporated into the thinking of governmental and non-governmental practitioners. Notably, the concept of translocality was employed, in a somewhat general sense, by the Cambodian government in the wake of the 2008 recession in order to assuage concerns over the narrow industrial base of the economy. As was argued at that time, 'those who lose jobs in the manufacturing sector can return to farming' (Un and So, 2009). Yet not only is this claim inaccurate—contemporary Cambodian agriculture is increasingly supported by modern sector wages, not separate from it—but the corollary to this flexibility: debts, obligations, and multi-local pressures remained largely unconsidered.

Only by considering both sides of this equation is it possible to understand how 'patterns of inequality persist' despite Cambodia's 'impressive economic growth and change (Deth and Bultmann, 2016: 88). The same linkages, obligations and systems of reciprocity that facilitate translocal livelihoods also sustain inequality, not in spite of, but due to households' and families' ability to share resources, money, and labour. Macroeconomic inequality figures therefore exhibit a misplaced focus on large scale trends, where inequality in highly mobile societies such as Cambodia is engendered often at the smallest scale. Moreover, as the following sections will highlight, this is a phenomenon with far more than merely economic roots, but one manifesting in complex ways across narrative, as well as financial, spheres of value.

2.2 Physical Capital and the Translocal Narrative of Land Transfer

The financial flows engendered by translocal working practices are key to understanding why 'despite strong economic growth in the last decade, the inequality gap has been widening over time and there is no indication it will decline in the near future' (Kung, 2012: 224). However, this is not to argue for a purely monetary perspective on wealth distribution. As a raft of large-scale surveys have highlighted 'Cambodia's weak formal institutional structures' (Hill and Menon, 2013: 50) and public perceptions of economic inequity are directed with growing regularity towards institutional corruption and government mismanagement. As a result of these shortcomings, land—that most immobile of assets—has flowed inexorably from the poor to the rich of the Kingdom.

The extent and rapidity of this transfer is rendered all the more alarming by historical context. Cambodia's physical landholdings have emerged today from an unusual degree of equality in the 1980s, at which time each household was awarded a plot of land during the short lived *krom samaki* [solidarity group] collective farming arrangement. Although variations existed between regions— with land being allocated according to the number of household members in some cases and divided equally between households in others (Amakawa, 2007; Yagura, 2005)—land holdings during the PRK period may in all cases be 'presumed to be relatively equal' (Yagura, 2005: 30).

Today, by contrast, 'some 20–30% of the country's land has passed into the hands of less than 1% of the population' (USAID, 2011: 5) and median land distributions amount to considerably less than 1 Ha per household. Moreover, the country is characterized by high levels of landlessness—estimated at 40 per cent of all households in 2009 (GTZ, 2009)—an issue that points to growing reliance on non-farm income sources and hence heightened vulnerability. Thus:

> Combined with frequent distress sales, associated with food insecurity and high health care costs, the pattern of land ownership has been transformed since the socialist era ended in 1989. A relatively egalitarian distribution in 1989 has given way to sharp inequality.   (Hughes, 2008: 71)

The reasons that underpin these issues are complex. However, the linked processes of migration and microcredit (Bylander, 2015) have proved especially influential, combining to entrench and deepen existing inequalities through differences in how households acquire and utilize debt (Yagura, 2015; 2005). Above all, these have combined to monetize and mechanize the rural economy, rendering smallholder farming—especially in the current high risk environment for agriculture (Bylander, 2015, 2014; Oeur et al., 2012)—increasingly dependent on modern sector wages and credit to purchase farming inputs whose prices have risen rapidly in recent years (Sothearith and Sovannarith, 2009). Environmental shocks therefore force larger debt repayments, engendering higher levels of remittances and more migration to compensate, up to the point that a household's only remaining repayment mechanism is land sale to less vulnerable households.

Moreover, whilst market and environmental forces have played a key role in engendering land inequality in recent years, these structural changes have not occurred in a social vacuum. Rather, 'landholding is intertwined with poverty in rural Cambodia' and its redistribution is therefore linked into numerous structural dimensions of wealth and its absence (Yagura, 2015: 173). This applies not only to the losses experienced by smallholding farmers, but also to the Kingdom's larger scale and state mediated land transfers, whose abuse has seen 'eviction...become a feature of Cambodia's development', as the lines between private, public and state property have been blurred by competing or unclear legal frameworks (Hughes, 2008: 71).

Two issues in particular have seen smallholders divested in large numbers of their farmland: the lack of formal titling throughout the country, and a constitution which favours the state in cases of marginal ownership. Though often considered separately, both concerns are closely interrelated (Dwyer, 2015), have strong linkages to wealth inequalities, and prejudice the rights of the worst off to the greatest extent. Indeed, not only is a land title 'difficult and expensive to come by' (Hughes, 2008: 71), meaning that the majority of those who possess one are from better off socio-economic strata (Bulgaski and Pred, 2010), but the contemporary problem of land concentration that underpins these rural inequalities exists to a large extent as a 'consequence of the governmental land concessions and land purchases' (Un and So, 2009: 129).

That such a vast project of national land sale has been conceivable, let alone workable, owes much to national narratives of property and settlement. Widely held perceptions that 'peripheral areas have large swaths of "under-populated" and "available" land' (Dwyer, 2015: 910) have their roots in deeply entrenched cultural dichotomies between rural/urban and village/forest (Chandler, 2009) which facilitate state actions beyond legal frameworks. Viewed thus, the contemporary system of land ownership, wherein land that lays fallow 'automatically falls into the state's possession' (Schneider, 2011: 13) and 'the ambiguous nature of state land and the convenient transferability of state public land (such as forests, fallow, or non-private lands) to state private land' (Schneider, 2011: 15) are merely two dimensions of a more generalized willingness to exploit and ignore popular conceptions of land rights. As a result, 'the rural poor are left out of the picture because of the power disparities in the system of land governance' (Schneider, 2011: 15), which have themselves led to:

(1) a large gap between the country's legal framework and the implementation of the country's land concession policies and (2) a complete disregard of the country's customary land rights. Widespread corruption and nepotism encourages growing inequality in land ownership and a significant power imbalance between small groups of powerful, politically and economically well-connected elites and poor and vulnerable people in Cambodia.

(Oldenburg and Neef, 2014: 49)

Rather than ameliorating these, the implantation of Cambodia's nationwide land titling program, the Land Management and Administration Project [LMAP], appears to have 'effectively formalized, and arguably deepened, structural inequality in land tenure and administration in Cambodia' (Bulgaski and Pred, 2010: 9), functioning less as a solution and more as a fig leaf to large scale reallocation and redistribution of land holdings across the country (Dwyer, 2015). The growing inequality of land distribution evidenced in Cambodia therefore exemplifies not only opportunism, but the structural inequities associated with what Dwyer (2015: 903) calls 'the formalisation fix', or 'the proposition that property formalization constitutes a preferable front-line defense against land grabbing'.

Thus, despite their origins in cultural narratives of land rights, these issues do not begin and end with domestic governance, but in local 'articulations' (Springer, 2011: 2554) of global political-economic policy. Local institutions' allocative processes are 'non-transparent' and 'deeply corrupt' in many cases (Hill and Menon, 2013: 51), but the changes underway in Cambodia's land distribution over the last three decades are as much a result of Cambodia's position within the world as the rationality of its local institutions. Cambodia is 'a small, open, fragile, post-conflict economy' (Hill and Menon, 2014: 1) whose society has been buffeted by wide ranging neoliberal reforms in a number of ways (Louth, 2015).

Land concessions—small and large, forced and 'voluntary'—have proved especially influential in this respect, creating a pervasive atmosphere of tenure insecurity that has become intimately linked both to the proliferation of microcredit and the advent of modern sector work. Viewed thus, the redistribution of land towards larger landholders reflects a broader process of mobility, wherein human and economic flows are predicated on instability as much as opportunity. Translocality, not only of capital and labour, but also narrative, has been central to this process. Whilst ties to the land are being loosened by the obligations and opportunities of the modern sector, conceptions of property have coalesced from various places and scales to engender an environment amenable to the large-scale transfer of physical wealth. Dynamism, thus, has been the servant of disadvantage.

2.3 Embodied Inequality: Translocality of Health and Education

Health in Cambodia has long been viewed as a matter of economic geography. As studies consistently demonstrate, 'inequality in health status and access to health care between rich and poor and between urban and rural residents is phenomenal' (Hong and Them, 2015: 1040) and geographical factors, intertwined with economic ones, remain foremost in shaping it. In part, these gaps are explicable by 'non-inclusive economic growth... particularly across wealth and geography markers' (Jimenez-Soto et al., 2014: 1). However, this explanatory framework risks missing much of the embodied character of Cambodian development. It remains rooted in the rich/poor and rural/urban dichotomies discussed above, leaving a significant amount of the multi-local, mobile and intra-locational nuance by which health inequalities manifest underexplored.

A particular issue is that a priori geographic categorizations—as well as ignoring that 'health inequality is mainly due to within-location inequality' rather than that which occurs *between* locations (Fujii, 2013: 343)—is their inherent analytical stasis. Rarely are the translocal linkages that transect geographical categorizations explored in their relation to health issues and whilst various studies have emerged concerning the health status of migrant workers (e.g. BFC, 2016; Enfants & Developpement, 2015, PSL, 2014), the outcomes they record are invariably viewed only in their urban context. Thus, despite strong evidence that the urban livelihoods of migrant workers are dynamically interlinked with rural circumstances (e.g. CARE, 2017; Parsons, 2016; Parsons and Lawreniuk, 2016), the impact of rural pressures, such as floods, droughts, and agricultural seasonality on health outcomes remains unexplored.

Categorical constraints such as these manifest not only horizontally, restricting health analyses to the use of a mono-locational lens, but also vertically. As alluded to in the recent literature on 'fleshy geopolitics' (Dixon, 2014: 2) and earlier by Baviskar (2001), health inequalities are often more starkly felt within the boundaries of social institutions—the home, the village, gender—than purely geographical ones. Gartrell (2010: 289) for instance, highlights how 'geographical processes fix disabled people in their socio-spatial place, which together with ideological and structural inequalities distinguish and entrench their poverty from that of other social groups' (see also Gartrell and Hoban, 2016, 2013). Embodied inequalities, these studies show, are neither simple nor static in their manifestation, but emerge from a complex milieu of intersecting, multi-scalar factors.

The translocal nature of deprivation in Cambodia means that these factors often originate in spatially distant locations. Urban rents and conflicts between factory unions combine with flooding, insects, or the dynamics of irrigation on the family farm to engender specific health outcomes in individuals. Workers complain of headaches and fatigue, as low-quality rice takes its toll on the body during harvest periods. However, these ailments are patterned by culture and narrative, finding form and direction through the social relations in which migrant livelihoods are embedded.

Gender, in particular, plays a role. Men who lack the social connections to find work in the best sites suffer skin and lung problems due to the materials they work with. Women in similar situations endure additional complications to their reproductive and maternal health, as an inability to pay for medicine and consultations hinders their ability to care for themselves before and during pregnancy (Enfants & Developpement, 2015; PSL, 2014). In both cases, though, translocal livelihoods are a key predictor of health outcomes. Not only are the best off less burdened by remittance commitments, but they are more able, also, to build the connections they need to find work in the most desirable worksites with the best healthcare facilities. As a result, they and even their children embody their geopolitical disadvantage, as subtle distinctions in diverse locations are rendered permanent in flesh.

Moreover, embodiment in this sense does not end with the body itself. Critical perspectives such as these may be applied with equal efficacy to the other major segment of human capital: education, where 'stark inequalities are visible due to the legacy of a violent 30-year period [during which] much of the highly skilled human capital stock was killed or fled from the country' (Collins [in Holsinger and Jabob, 2008]: 191). The peculiarity of these circumstances mean that, as with its recent land distribution, Cambodia is 'an extreme but particularly illustrative case' (Dwyer, 2015: 905) in that inequalities are laid bare by highly visible gaps in educational attainment across a number of intersecting axes.

Adopting a livelihoods perspective on issues of this sort is rendered all the more vital by the 'fuzzy distinctions between government systems, markets, and the formal education system' in Cambodia (Dawson, 2010: 20). Students are expected to purchase many of their own materials and the widespread practice of teachers soliciting regular payments from their pupils (Dawson, 2010), reflect salaries in the sector of between $200 and $300 per month, depending on the level of teaching (Ros, 2016). Although less prevalent in rural areas than urban ones (Brehm, 2016), this direct financial burden serves both to exclude children from lower income families and to inhibit the quality of their education by restricting their access to key resources.

As modern sector wages provide additional capital for education, these practices are gaining ground rather than retreating. In recent years, 'hidden privatization has entered public schools in the form of private tutoring' (Brehm and Silova, 2014: 95) to the extent that in some cases 'teachers withhold curriculum content in order to force students to pay them for private tutoring lessons' (Dawson, 2010: 21). This growing trend has engendered a robust quantitative relationship between private tuition attendance and exam scores (Brehm and Silova, 2014: 94) and thus strengthened the existing relationship between educational attainment and wealth. In this way, the 'fuzzy distinctions' in the narrative of education between state and private activity (Dawson, 2010: 21) coalesce into clear cut and durable divisions, as educational arrangements 'stratify Cambodian youth along socioeconomic lines' (Brehm and Silova, 2014: 94).

In particular, it is observed that 'female, rural and minority groups [are] being disproportionately disadvantaged' by systems which fail to account for the specific pressures they experience (Collins [in Holsinger and Jabob, 2008]: 190). However, in the rapidly developing context of Cambodia, these points of vulnerability paint only a broad brush picture. In reality, 'cross-cutting income disparity' and variation 'within a locality' (Hamano, 2010: 4) point to the intersection of multiple factors in generating educational iniquities. Thus, whilst institutional weakness, corruption, and chronic underfunding contribute the underlying context of Cambodian education, they offer little insight into the specific landscape of access within this. That rural areas have poorer educational outcomes is known, for instance, but how the experience of petty traders' versus farmers' or fishers' schooling outcomes differs far less well understood.

Changes to each of these livelihoods have driven new patterns of rural and urban mobility whose specific impact on education is profound. As hybrid, rural–urban arrangements for income generation continue to spread, not only waged, but also non-waged migrations have become crucial to translocal households and their members (Lawreniuk and Parsons, 2016). Children are now more mobile than ever, frequently dividing the year between multiple places as their parents seek out wage labour or modern sector salaries. In some cases, this means missing significant portions of the academic year, whilst either helping parents to undertake migrant work in difficult and often dangerous locales, or simply accompanying them for lack of care elsewhere.

Even amongst those who do not migrate with their parents, grandparental care is often limited by physical incapacity, lack of funds (UNICEF, 2017) or the need for these carers to take on additional household and agricultural roles at the same time (Lawreniuk and Parsons, 2016). Older family members are increasingly taking on crucial supportive roles in relation to translocal livelihoods, yet the range of new duties involved is overburdening in some cases. Many grandparents struggle to meet a household workload that previously fell to a whole family unit on their own, leaving little time for the everyday intricacies of childcare. Children therefore often find themselves largely unsupervised during their parents' absence, left to their own devices without encouragement or incentive to pursue their studies or even attend lessons in many cases.

Consequently, the migration of parents 'is found to have a significant negative effect on school attendance', leading to sporadic attendance and earlier dropout (Hing, Lun, and Phann, 2014: 1). Moreover, whilst direct costs such as these are a key factor linking educational to economic inequality, indirect and opportunity costs also play a vital role. Like health inequities, these are best viewed through the lens of social institutions, norms and livelihoods, without which perspective complex intersectional disadvantages are frequently missed. In particular, local issues have long played a role here, with periods of seasonal activity often impacting on schooling outcomes via missed days, as children are needed in economically productive roles (Stirbu et al., 2010).

Thus, it is mobile and multi-local labour patterns that are increasingly defining the educational landscape. As livelihoods have diversified during the past two decades, such strategies have not only had a significant impact on the overall economic status of a given household, but decision making within it also. It has long been noted, for instance, that a significant proportion of garment workers are taken out of school in order to work (Sim, 2004) and this is an issue which extends to multiple forms of mobility, as household economic factors tend to be the first reasons given by children dropping out of school (USAID, 2013).

As McKay et al. (2016) note, these differential trends in education feed directly into migrant labour markets: education is positively correlated with employment in a formal sector job, whilst poverty exhibits a negative correlation. However, the high proportion of remitted funds spent on education (Hing et al., 2014) further complicate matters, suggesting as they do that education may be sacrificed by one generation as part of a deliberate longer term strategy to acquire higher levels of education for the next. Whether or not such a strategy is successful depends on a complex combination of factors, from rural agriculture to urban industrial action, each of which is itself linked into the other.

Like health inequalities, then, differences in educational attainment constitute the embodied manifestation of multi-scalar systems of mobility rooted in norms, institutions and global geopolitics. Far from being rooted in static geographic classifications, these inequalities may be fungibly transduced from one dimension of poverty to another, cross-cutting occupational and geographic boundaries as they manifest in the everyday experience of translocal livelihoods. Inequality, in this sense, is therefore not durably embedded, but constantly refreshed and sustained by systematic phenomena, manifesting as mobile and multi-local even where it adopts the most seemingly permanent of forms. Yet conventional measures do not adequately reflect this, suggesting instead an abstracted and compartmentalized view, ill-suited to livelihoods, hierarchies and prejudices that are lived across more than one place. What is necessary is a mobile conception of inequality.

2.4 Conclusion

It has long been noted that poverty is multi-dimensional (ADB, 2014), manifesting in numerous forms of capital and best understood using a variety of indicators. Similarly, the distribution of poverty—or more broadly, inequality—is recognized to be geographically determined, whether viewed in terms of urbanization, agro-ecology, or distance from key infrastructures and services. However, despite broad acceptance of the complex inter-linkage of these factors, analyses have thus far been hamstrung by their basis in economic conceptualizations and the pervasive assumption of stasis.

In reality, inequality in Cambodia is mobile—often hyper-mobile—and manifests in multiple places and numerous forms simultaneously. It is both rural and urban, local and multi-local, embodied and external. However, traditional measures of inequality, which are almost invariably based on national scale surveys utilizing economically biased and inherently static livelihoods metrics to categorize the population, have failed to capture this complexity. Instead, what is measured is a partial inequality, effective at detecting static forms of deprivation, but inadequate to exploring wealth distribution in a mobile context.

In seeking to rectify this, Cambodia is ideally suited to the investigation of mobile inequality. Its relatively small size hastens and accentuates the flow of people from one locality to the next, strengthening economic and social inter-linkages and enhancing the dynamism with which shocks and stresses originating in one place are felt elsewhere. At the same time, the Kingdom's highly open economy and unstable environment for agriculture have catalysed movement across the country, as rapidly changing rural and urban economies push people both to enter the modern sector and retain a foothold in the traditional one. This context enhances and clarifies inequality, laying bare its mechanisms and highlighting its interaction with social institutions and norms.

As a result, two characteristics in particular emerge. First, the duality of inequality; how physical and embodied inequities work in combination to influence the character and outcomes of movement. Secondly, the translocal nature of these iniquities, or how capital imbalances originating in one place manifest between and across boundaries, linking and segregating people in a manner that transcends spatial logic. Nevertheless, these diverse, complex and dispersed strands are drawn together through their persistent connection to multi-scalar systems of social organization. Inequality is therefore neither atomistic in manifestation, nor independent in form, but operates according to pre-existing normative and structural frameworks within society. The intersecting role of each must be considered in order to produce an effective analysis of unequal livelihoods in an increasingly mobile and globalized world.

3

Mobile Inequality

Embedding Economic Flows in Mobile Social Structures

In Cambodia, narratives of change abound. Amongst the land concessions to the North of the country, they have sometimes meant resistance, as large companies are cast as villains against whom village vanguards appeal to the government as a child to a parent (Parnell, 2015), whilst elsewhere they have encouraged acquiescence, as individuals and communities draw upon cultural values of endurance. Yet the narrative that has dominated Cambodia's economic development in recent decades is an imported one: the broadly conceived notion—often attributed to John F. Kennedy, but used with increasing frequency to justify laissez faire economic ideology—that 'a rising tide lifts all boats'.

For many years, the evidence appeared to support this account. During the breakneck growth years of 2000 to 2008, every boat really did appear to be rising, albeit at vastly different rates in some cases. Nevertheless, it is a narrative that flatters to deceive. As set out in Chapter 2, viewing Cambodia's growth and the inequality it engenders through a macroscopic lens elides key features of its incidence. Subtle differences matter greatly here and one of the gravest errors made by analysts is to present the big picture without attending to these details. In Cambodia, farming is not just farming and garment work is not just garment work. Inequality pervades and links both sectors, turning a difference of degree into one of kind. After all, anybody can find themselves in a difficult or uncomfortable situation. What matters is the ability to adapt or abandon it. The better off do not put up with what they cannot stand, whilst the worst off must do exactly that: bearing conditions that no one else will, out of necessity and isolation.

Everyday choices have major implications in this respect. Migrants from the poorest households make substantial 'sacrifices' in the urban space (Lindley, 2009: 1326), eating low-quality food, replenishing their clothing rarely, and spending little on attending social events. By contrast, those migrants upon whom remittances and rural dependency are less burdensome enjoy more of the fruits of their income, consuming goods and services and, crucially, becoming more socially involved with other migrants. The linkages produced by social behaviour such as this appear to be one of the means by which wealthier migrants gain access to

Going Nowhere Fast: Mobile Inequality in the Age of Translocality. Sabina Lawreniuk and Laurie Parsons,
Oxford University Press (2020). © Sabina Lawreniuk and Laurie Parsons.
DOI: 10.1093/oso/9780198859505.001.0001

urban opportunity (Yu and Berryman, 1996), thereby entrenching and deepening patterns of rural inequality in the long term.

Focussing on three neighbouring villages in Krang Youv commune—a development test case famously patronized by Prime Minister Hun Sen (Chim et al., 1998)—and the diaspora of migrants who emerge from there, this chapter constitutes an effort to better understand the interaction of social ties such as these with rural and urban society more generally. It does so by employing an eigenvector-based model of social network analysis to explore migrants' and their household's social relations within the context of their migrations and household economies, in order to explore how social ties derive from and underpin migrant inequality.

Having grounded the study in its empirical and conceptual context, the chapter presents data on this theme in three parts, the first of which emphasizes the intra-occupational nature of migrant inequality and how remittances generate and maintain that inequality. Secondly, it will outline the associated network dimension of this relationship: how migrants from rural households with differing incomes behave distinctly in the city and how households which depend more or less on their migrants behave differently in turn. Finally, the inter-relationship of social networks, remittances and livelihoods 'advancement' in the urban space will be explored.

3.1 Mobile Inequality in Contemporary Cambodia: Structure, Gender and Culture

The dominance of Cambodia's economy by key modern sector industries, coupled with the 'demographic dividend' of a young population entering working age (Beyene, 2015) mean that Cambodia is increasingly a country 'on the move' (MOP, 2012: 1). Moreover, it is moving overwhelmingly in one direction. The primacy of Phnom Penh as an economic centre (MOP, 2012; Sheng, 2012) has seen it balloon in size in recent years, doubling in size between 2004 and 2012 and now accounting for 1 in 10 Cambodians, where in 1998 the figure stood at 1 in 20 (MOP, 2012).

However, despite the macroscopic flow of people from the countryside to urban areas, this process should not be viewed as unidirectional. Rather, this phenomenon of human movement is made up largely of far smaller cycles lasting from a mean few weeks in the case of begging migrants (Parsons and Lawreniuk, 2016a) to a few months or years in the case of construction and garment workers (ibid.).

Furthermore, that the average duration of stay in certain migrant dominated areas appears to have declined significantly during the past half-decade (Parsons and Lawreniuk, 2017) highlights the extent to which rural and urban areas have become interconnected by these complex systems of movement and remittance.

They are linked not only by economic imperatives, but also changing patterns of social relations (Kheam and Treleaven, 2013; MOP, 2012; Lim, 2007) and cultural norms (Parsons and Lawreniuk, 2017; Czymoniewicz-Klippel, 2013; Elmhirst, 2007) generated in part by the preference for female workers within the dominant garment industry (Cuyvers et al., 2009).

Indeed, as various authors have noted (e.g. Bylander, 2015; Brickell and Chant, 2010) young women are becoming increasingly important economic agents in rural villages—with both positive and negative implications for their personal livelihoods—whilst structural unemployment amongst young men has been linked to the emergence of groups of gangsters in rural and urban villages (Czymoniewicz-Klippel, 2013; Elmhirst, 2007; Silvey, 2001). These shifting economic norms are negotiating new narratives of gender in Cambodia, though often without loosening their restrictive characteristics (Derks, 2008).

Similar processes are examined in some depth in the household literature on migration (e.g. Rindfuss et al., 2012; Resurreccion, 2005; Silvey, 2001; Kabeer, 1997), which explores how differing local circumstances interact with household dynamics to produce unique networks spanning rural and urban areas. Nevertheless, though the case made here is strong, norms and narratives of gender have yet to achieve full integration in migration models. When culture has been admitted to the migration literature, it has tended to be in the form of 'a residual value' (Levitt and Llamba-Nieves, 2011: 2) in the migration process, highlighting 'the need for a more complete theory of migration that incorporates notions of cultural dynamics as they relate to behaviour and societal outcomes' (Curran and Saguy, 2013: 54).

This shortcoming manifests in a number of ways, but is felt especially acutely in relation to gender norms, the feedbacks and interrelationships of which have played a defining role in the development of the Cambodian migration system. As Tables 3.1 and 3.2—included here for comparative purposes due to the small number of migrants from Krang Youv working in certain occupations—show, for instance, both migration from Krang Youv specifically and to large migrant enclaves such as Teuk Thla more generally, is dominated by women. However, female migrants are not equally represented across occupations. Rather, gender

Table 3.1 Gender ratios of major occupations in the study

Occupation (N=50)	Male (%) (12)	Female (%) (38)
Garment workers (N=22)	13.6	86.4
Market traders (N=8)	12.5	87.5
Construction workers (N=5)	12.5	87.5
Mechanics (N=3)	100	0
All Krang Youv migrants (N=50)	38	62

Table 3.2 Comparative data from large-scale study on Teuk Thla (Parsons, Lawreniuk, and Pilgrim, 2014)

Occupation (N = 1642)	Male (%)	Female (%)
Garment workers	89.7	10.3
Market traders	14.3	85.7
Construction workers	95.5	4.5
Mechanics	100	0
All Teuk Thla migrants	30.8	69.2

Table 3.3 Comparing characteristics of migrant and non-migrant households in Krang Youv

	Households with at least one migrant (N=61)	Households with no migrants (N=90)
Percentage of total	40.4	59.6
Median rice land area	0.5 Ha	0.06 Ha
Median income (including remittances)	$3068	$900
Percentage of income from remittances	43.5%	0%

biases associated with certain jobs appear to strongly constrain migrants' opportunities and therefore both the incidence and form of their migration.

Similarly, household socioeconomic factors play a key role in the migration process. Rather than constituting merely an addition to household livelihoods portfolios, the declining viability of traditional agriculture due to climactic uncertainty and rising input costs (Bylander, 2015, 2014; Oeur et al., 2012) has placed migration at the core of household strategies pursued by rich and poor alike.

Indeed, as Table 3.3 demonstrates, households with migrants are on average richer than those without. Moreover, although remittances contribute substantially to this distinction, they do not account for it entirely. Rather, the significant difference in assets and income suggests either that migration aids in the acquisition of rural assets, or conversely that their possession aids in the successful pursuit of migration.

As the social network perspective adopted in this chapter will show, the reality is both: poorer households in many cases lack the networks necessary to undertake migration in the first place, but this inequality of opportunity is not limited to the initial migration event. Rather, even if a member is able to achieve urban employment, poorer households tend to be impeded in various ways from accessing the long-term benefits of their labour. Rural assets and income are hugely important in this respect and impact not only upon the long- term prospects of migrant households, but of migrants themselves, whose experience of migration differs markedly as a result of them.

The remainder of this chapter aims to demonstrate the mechanisms and patterns by which these inequalities become mobile and multi-local. It presents data from three primary fieldsites researched between March and August 2013: the rural commune of Krang Youv in Kandal province close to the Southwestern border with Takeo; the peri-urban garment worker enclave of Setbo, to the south of Phnom Penh; and in Phnom Penh itself. Across all three locations, 151 informants were selected via a network mapping methodology originating from three initial interviewees, each sited in one of three areas identified during a period of preliminary qualitative fieldwork as being respectively the richer, medium, and poorer parts of the local area. A final phase of research sought to 'track out' of migrants to their urban and peri-urban worksites by meeting and interviewing migrants originating from the sampled network of households to produce a base sample of 50 urban informants.

3.2 Financial Inequality

3.2.1 Conceptualizing Migrant Inequality

In conceptualizing Cambodian migrant livelihoods it must be recognized that intra-occupational differences are often as important as inter-occupational ones. The conditions faced by construction workers, for instance, differ markedly depending on whether they work for a large company, a single foreman, or anything in between; whether they live with family in a rented room, or onsite for months on end. Their wages too depend upon a number of factors. For instance, many female construction workers deemed weaker and hence less valuable than their male counterparts despite undertaking the same labours, work for 7–8000 riel [$1.75–$2] per day, alongside unskilled men at 10,000–12,000 riel [$2.50–$3], and skilled workers earning up to 20,000 riel [$5].

Similarly, although garment workers' salaries tend to be relatively homogenous within a given factory (excluding new starters and line leaders who make up a small proportion of most workforces), small variations in wages, conditions and terms are seen as hugely important by those affected. The provision by a factory of a single free meal per day, or slightly greater flexibility in requesting time off to assist the rural household with agriculture, can mean the difference between merely subsisting, or saving for a future business. Such benefits are jealously sought and many factory workers change jobs multiple times in pursuit of the best combination of conditions for their circumstances.

Differences such as these are easily to overlook, but they are nevertheless a key base of inequality in migrant enclaves, where income inequality is relatively low compared with the city as a whole. However, it is not earnings and expenses alone which determine migrant livelihoods, but these factors in combination with outgoings from remittances. Salaries and familial claims upon them cannot be separated for most migrants: it is net income, inclusive of rural–urban flows of food and urban–rural flows of money, which underpins the reproduction of rural structural inequalities in the urban space.

3.2.2 Remittances as a Factor in Migrant Livelihoods

The case of Cambodia—where over a third of Cambodians have now lived in more than one province in their lifetime, with almost 10% of the population living abroad, predominantly for work (UNESCO, 2018)—provides an exemplary test case for the growing recognition within the migration literature (Coulter, Ham, and Findlay, 2015; Jensen, 2009; Law, 1999) that narratives of migration are dynamically and bi-directionally linked to its economic dimension. So vital are remittances becoming to Cambodian households, for example, that narratives of home, gender and duty are being actively recast in the light of new economic imperatives (Brickell and Chant, 2010). A key example of this is the changing role of women and girls in view of their perceived superiority as remitters. As many now say: 'girls are the only ones who can really help the household' (Village Chief, Prey Veng, Parsons, Lawreniuk, and Pilgrim, 2014: 16), a sentiment with vast consequences for power relations in the household and beyond.

Moreover, such statements are partially supported by the data, which indicate that female migrants do indeed remit more on average than their male counterparts. Yet as indicated in Table 3.4, the proportion remitted is almost exactly the same for both groups. This suggests that whilst there is some truth to gendered remittance narrative, it is underlain not by the widely perceived idea that young men are more likely to waste their money drinking and fighting (shopkeeper and rented room owner, 21/06/13), but by lower wages (when viewed on a monthly or yearly, rather than daily basis) derivable from the migrant employments available to the majority of men.

Table 3.4 Data on remittances by gender (N=50)

	Male	Female
Percentage of total remitters	30.3	69.7
Mean remittances per month ($)	37	45
Percentage remitted	35.2	35.5
Mean monthly wage ($)	118	129

Construction work, for instance—the major employer of male of migrants in Cambodia—though hugely varied in its pay and conditions (see Figure 3.1), most commonly hires employees on a per job basis, meaning that most construction workers expect to spend a period of the year without employment, whilst their foreman or boss awaits a new contract. By contrast, although job security in the garment industry is declining (Parsons and Lawreniuk, 2017; Arnold, 2013), it has in the past been sufficiently superior not only to impact on remittance patterns, but to generate feedbacks such as a preference on the part of Microfinance Institutions to agree loans to those with family members working in the factories (Krang Youv villagers, 03/06/13).

In spite of such narratives of distinction, however, both men and women in the major migrant occupations are highly efficient remitters. Indeed, as Table 3.5 shows, migrants remit an average of 36 per cent of their total monthly income (after overtime), a figure which includes the quarter of migrants in the study who do not remit to their parental household. Amongst only those migrants who do remit, the figure is very close to half of gross income.

Thus, the average migrant in this sample retains only $62 per month, or $2 per day for their own subsistence, of which workers generally report that each of their three meals costs an average 2000 riel ($0.50). In addition to this, their rent—which varies significantly due to the number of workers living in one room—amounts to a further $10–$15 per month: almost the entirety of what remains.

Figure 3.1 Day breaks over a construction site in Phnom Penh, 2018. The green corrugated blocks are temporary on-site accommodation for workers.
Source: Courtesy of Thomas Cristofoletti/Ruom/Blood Bricks/Royal Holloway.

Table 3.5 Absolute and relative remittance behaviour of selected occupations (n=50)

Occupation	Mean salary ($)	Mean remittances per month ($)	Mean percentage of salary remitted (%)
Garment workers (N=22)	127	61.30	49
Construction workers (N=5)	129	59.38	45
Market traders (N=8)	190	10.75	6
Mechanics (N=3)	312	20	12.5
Mean (across whole study) (N=50)	131	42	36
Mean for remitting migrants (N=33)	125	56	47

Taking the mean figures alone, then, migrant workers appear to subsist at the limits of their income, with almost nothing remaining in case of illness or other unforeseen expenses. Indeed, a lack of personal security resulting from such maximal remittances was a frequent refrain during interviews, as evidenced in the following testimony:

I don't make enough to send money home [easily]. Every month I can send only $70 or $80, which isn't enough for the people back home. It's never enough because my parents have a debt from paying for my mother's hospital treatment. She had something wrong with her throat and needed an operation to take it out. [I know some people but] we only meet each other at the factory every day. I never go anywhere to eat or drink anything because I need to save money and don't want to go anywhere I can spend it.   (Vireah, 13/08/13)

Vireah's testimony is emblematic of the second of the concerns to be addressed herein. As he explains, the compulsion to remit a large proportion of his salary leaves him with no disposable income, thereby preventing him from integrating into the migrant community in a meaningful manner. Similarly, Chea, a garment worker from Boeung Tumpun, Phnom Penh, explained that:

My father is already retired and so doesn't rent any livestock any more. He has divided all of his land between his children and lives only on what they send him. [Now though], it is only me sending money because the others are either married with families or studying... As a result, I rarely go out because I have to look after three people – my father and two sisters [who are studying]. Generally I don't go anywhere really.   (Chea, 22/07/13)

Although workers from poorer backgrounds such as these are physically present in the city, they remain conceptually annexed to their rural households. Yet these reports of urban isolation—which are just two amongst a great many—contrast markedly with those of some wealthier migrants, who expect not only a degree of freedom in their urban lives, but also the opportunity to invest. As Leap, a car mechanic, outlined, his situation is one characterized by financial self-determination, familial support and clear prospects for the future:

> I never send money to my parents because they say there is no need. They have enough money. I am saving for two things: getting married and to open a shop of my own. This will cost about twenty thousand dollars and with my parents help with half the cost, I will be able to do it next year.   (Leap, mechanic, 21/07/13)

Similarly, a female non-remitter commented that:

> My family at home generally don't need any money from me, because they have other children working too. They rarely ask for anything from me. I save the money I don't spend and now have about two months salary. I'm going to use it to buy jewellery and when I get married I will use this money to make more money so that I don't have to rely on my husband.
> (Mony, garment worker, 01/08/13)

Both of these testimonies highlight the ability the respective informants possess to save and invest their salaries. What is notable, though, is the difference in their urban occupations. Although the first of these informants is one of the highest earners in the sample, a car mechanic with a monthly salary exceeding $500 after 'tips', the second is a garment worker, earning a basic salary of $75 per month and only $115 including overtime.

Nevertheless, though their urban income may differ, it is their rural wealth that determines their urban lifestyles and aspirations. Despite many of them boasting no greater earning capacity than their counterparts from poorer rural households, migrants from wealthier families spend, as highlighted throughout the qualitative data, considerably more on leisure whilst resident in the urban space. Thus, those whose families depend less (or not at all) upon their remittances enjoy economic freedoms sufficient to mark them out as a financial elite in spite of the general parity of wages in some industries. In addition to enhanced leisure activities, such migrants are able to pursue a variety of investment strategies, most commonly either the purchase of 'gold, especially jewellery and rings' (Nakry, 01/08/13) or short-term money lending to migrants unable to subsist on their post-remittance salary between pay checks.

Moreover, this distinction is evident both within and between occupations. Not only do the data show a significant correlation (Spearman's = 0.491, p = 0.008) between migrants' urban occupations and their rural household wealth—in itself a vital source of segregation in an urban environment in which most social relationships are either familial or occupational—but also within occupations, where the qualitative data above and elsewhere emphasize that migrants cement relationships on the basis of the leisure activities they can afford to engage in.

Moreover, as the following sections demonstrate, the impact of familial wealth on migration is not measured in net income alone. Rather, the importance of familial networks in facilitating migration means that migrants remain tightly bound to the specific needs of their household throughout their migration, continuing to rely predominantly on family members for migratory information long after their initial arrival in the city.

3.3 The Dual Role of Networks: Facilitation and Obligation

3.3.1 Understanding Familial Networks

Despite the relative proximity of Krang Youv to key centres of industry such as Phnom Penh and Setbo, migration from the commune, as elsewhere in Cambodia, tends to be meticulously, albeit informally, planned and executed. Informal construction and garbage collection communities, for example, are strongly linked to sender villages, but even in the midst of much bigger and more established urban locales, migrants often comment upon the security they gain from small residential agglomerations involving a block or group of rooms. As one garment worker, resident in Phnom Penh, explained, 'I feel safe living here, despite living alone, because all of my neighbours in the rented rooms near me are from my home village' (Peou, 11/07/13).

Nevertheless, cohabitation with a family member is strongly preferred by migrants able to undertake it, with first time migrants moving in with a family member in over 98 per cent of cases (Parsons et al., 2014). Familial networks like these, often spanning rural and urban areas, are, as elsewhere in the global South (Chort et al., 2012), utilized as both information conduits and crucial logistical tools. Migrants rely on these networks throughout their migration, with initial migrations being undertaken without familial assistance in only 2% of cases, rising to 25% for subsequent moves.

Tables 3.6 and 3.7 therefore highlight two things. First, that migrants rely upon a wide variety of people to facilitate labour movement and second, that if the data are aggregated into moves achieved via social and familial linkages, then the latter appear to be the more widely utilized. Indeed, with familial help—including the extended household so as to include uncles, aunts and cousins—being the primary source of assistance in nearly 50 per cent of all moves, a sturdy case may be made for viewing migratory information as a predominantly familial resource.

Furthermore, the qualitative data strongly support this conclusion, including as they do numerous examples of migrants accepting family members' assistance as a

Table 3.6 Assisters[2] of migration (to and within the city) and their prevalence (N=50)

Primary source of information and/or assistance in labour movement	Percentage of total
Uncle/aunt	18.6
Friend from home village	18.8
Friend met in the city	4.9
Neighbour in the city	0.1
Cousin	9.4
Brother/sister	12.4
Husband	2.1
Parent	3.1
General information	5.1
Factory moved location	1.1
Self	24.6

Table 3.7 Simplified typology of migration assistance (N=50)

Primary source of information and/or assistance in labour movement (simplified)	Percentage of total
Family	47.5
Social	26.8
None	26.7

means to switch between places and occupations. Linkages of this sort appear to increase the speed and reduce the risk of movement to the extent that migrants become freer to pursue alternative earning strategies according not only to their households' but also their own needs and preferences. For instance, as highlighted by one such worker from the Krang Youv study site, who had left a well paid job in Thailand to work closer to her family, such advantages are prized highly enough to induce movement in pursuit of them:

> After a year and a half I felt like I had to return as I missed my parents and family too much. Once I got back, I decided to work in Andaing... [where] ... I receive only a medium wage, but live close to my brother and sister and can visit my parents on weekends or on holidays.   (Heng, 30/07/13)

Moreover, given that around 80 per cent of Cambodian migrants provide labour to the rural household at least once a year (Parsons et al., 2014), proximity to

[2] Assistance here has been self-defined by migrants, in response to the question 'who helped you most with your migration?' Follow up questions, however, reveal that this assistance always took one of three forms: information about the migrants' destination or a specific job, accompanying the migrant on their journey, or the provision of accommodation in the migrant destination.

home is a question of practicality as much as pleasure. For female migrants in particular (Elmhirst, 2002), the ability to participate in household seasonal labour is viewed not only as a financial, but also a moral issue. Yet despite such obligations frequently clashing with the ability to participate fully in migrant work, those occasional migrants who neglect to remit regularly are branded 'lazy' and 'uncaring' by their parents (Srawoon, 08/06/13). As a result, migrants whose households require both remittances and rural labour are often placed under extreme pressure to find arrangements which can accommodate both needs, influencing their urban behaviour in the short, medium and longer term.

3.2.2 Inequality of Obligation: Understanding Different Rates of Urban Movement

As the data above demonstrate, the rural household is both a facilitator and determinant of migrant movement. Moreover, it performs these roles both at the point of the initial move and once a migrant is established in the city. This is evident in a number of ways, with migrants frequently citing changes in their rural households' circumstances, especially serious illnesses, climate shocks, and failures in business, as prominent factors in their movements. For instance, as the garment worker daughter of an elderly widow related:

> [I came here because] I have debts to pay for the medical treatment of my father, who has now died, as well as my mother, who is currently ill and also requires hospital treatment. In total, this amounts to about $60 per month in interest only debt repayments...Any extra money I make goes to the home village. For instance, after overtime I may make $200 this month, so I will send it all home.
>
> (Tevy, 02/07/13)

Aside from medical treatment, failed ventures or investments undertaken by better off households are another common motivator of migration, as in the case of Lida, whose family had previously owned three hectares of rice land but had to sell it. She attributed this to bad luck and a lack of business acumen on the part of her father. In her words:

> We had to sell our land because my father failed in business. Previously he raised ducks, owning up to a thousand at one point, but they all got sick and died. When he started again, the same thing happened, so we had to sell our rice land. After the ducks [and with the remaining money from the land sale], we bought a large threshing machine for around $12 000, but it continually broke down and was expensive to repair. After a while we couldn't afford to keep repairing it and had to sell it for a cheap price...Now we have a debt of $4000, so I came here to work and send money home. (Lida, 13/08/13)

In spite of the many and varied events which might influence the decision to migrate, however, a common theme is that the movements and activities of poorer migrants are more closely determined by the household than those of their wealthier counterparts. Indeed, in spite of the variety of factors influencing migration decisions, a clear trend is visible between the rate of movement amongst migrant occupations, remittance behaviour, and household wealth.

These data indicate a negative relationship between total rural household income and the mean number of occupational changes undertaken by migrants since departing their village, shown in Figure 3.2. Moreover, as Figure 3.3 highlights, there is a negative correlation also between the rate of occupational movement and the percentage remitted by migrants (Spearman's = 0.359, $p=0.023$). This suggests, firstly, that migrants from poorer households tend to undertake a greater total number of moves than their wealthier counterparts and second, that the rate of such movement is associated with the need to remit larger sums.

One interpretation of this, supported by a significant correlation (Spearman's = -0.39, $p=0.012$) between net wages (gross monthly income—monthly remittances) and inter-occupational movement, is that even lower paid work may be deemed sufficient if a migrant is able to keep their entire wage, whilst a heavy remittance burden is likely to result in a protracted search for higher wages and/or more flexible conditions in order to satisfy the needs of both individual and household. As one such garment worker explained:

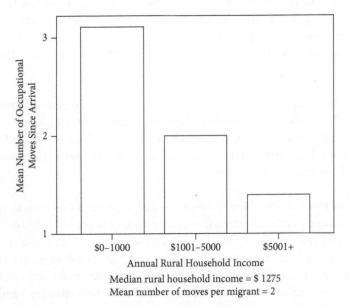

Figure 3.2 Household income as a factor in occupational movement (n=50)

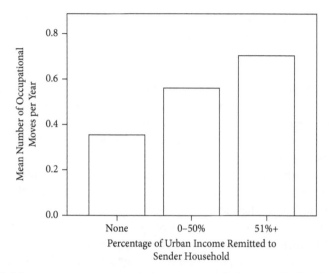

Figure 3.3 Mean moves per year against percentage of salary remitted to household (n = 50)

I left the first factory to return home because I got sick. Then I heard about a nearby factory opening. However, that factory only opened for seven months each year. Because my family was having a difficult time [with her father's illness] my mother asked me to change to a factory where I could work all year. [I did that but] a year later I had to follow my cousin to Setbo to another factory because the last one only worked eight hours per day so my basic salary was only $50–$60 per month. I needed more money for my family, so I had to come to Setbo.   (Reach, 09/08/13)

Nevertheless, although clearly influential in their own right, analysing urban activity in terms of remittances alone fails to capture the secondary impact of familial and social factors on migrants' structural integration into their communities. Seeking to account for this, the eigenvector measure of network centrality[1] has been utilized – following the method developed by Bonacich (2007) and Bonacich and Lloyd (2001) – as a means of measuring each household's and each migrant's level of social integration. This approach highlights a clear correlation between migrants' social integration into their urban environment and both total annual remittances received (Spearman's = 0.549, p = 0.01) and the percentage of annual income made up of remittances (Spearman's = 0.66, p = 0.04).

[1] Eigenvector centrality is a score that represents both the number of a node's (i.e., a person's) own linkages and the number of linkages possessed by those to whom they are linked. In this way, a connection to a node with, for example, 10 linkages of its own, is worth more than a connection to a node with only one, making the eigenvector a generally effective measure of social network centrality.

This may be interpreted in several ways. Firstly, more socially integrated households may allow themselves to become more dependent on remittances because their rural social networks afford them a degree of security in case of job loss. Alternatively, more remittance-dependent households may generate stronger social networks in the village as an information or loan generating safety net. In either scenario, although the causation is not obvious, the strength of correlation suggests that the social behaviour of rural households, and by extension the social structures in which they are embedded, are tightly bound to the circumstances of their migrant members.

This relationship further highlights the need for a more purposive interpretation of the role of social networks in the migration process than that which is often encountered in the literature (Palloni et al., 2001; Massey, 1990), which posits a fairly direct relationship between network centrality and wealth not supported in these data. Rather, given the importance of familial information and assistance to the migration process, as indicated in Tables 3.6 and 3.7, a preferable interpretation might argue that better connected households are able both to achieve a higher remittance return on migration, and that—due to the risk mitigating aspect of greater information access—they feel more able to depend upon it than the less well connected.

Moreover, it is not only household social networks which appear to influence and be influenced by the migration process. Combined with the percentage of salary remitted, migrants' own eigenvector centrality produces a model with an r-square value of 0.496 (p eigenvector = 0.005, p remittances = 0.007) suggesting that, as indicated by the qualitative data above, access to networks of information and the extent to which the sender household remains dependent upon the migrant's urban wages are both major factors in the migrant's urban experience. What remains to be demonstrated, though, is how these economic and informational factors combine to produce differential outcomes in the migration process.

3.4 Remittances as a Constraint on Urban Development

3.4.1 Rural Household Income as a Factor in Developing Urban Networks

As with capital and information flows, the role of migration in structuring social behaviour runs in two directions. Consequently, the relationship between social integration and remittance dependency may be found, in inverted form, in the urban sphere. There, migrants from wealthier households tend to be more central to their urban networks, as evidenced by the relationship between their eigenvector scores and their rural household income shown in Figure 3.4. In light of the

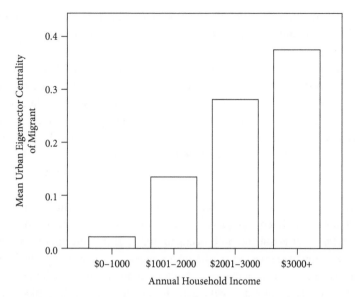

Figure 3.4 Mean remittance burdens of advancing and non-advancing migrants (n = 50)

testimonies above, this is likely to be due to the reduced burden of remittance payments imposed upon wealthier migrants, who are consequently able to use their salaries to engage in a greater degree of social activity in the city.

Rather than merely improving the experience of migration, however, the social relationships referred to above serve a vital purpose. The linkages they create between individuals who are relatively wealthy, or possess relatively high education and skill levels, are an invaluable resource in the transmission of opportunity and advantage, both for their constituent members and their families. Across networks such as these, apprenticeships are shared, money is invested, and information 'about business' (Chan, 24/06/13), money, and the urban environment, circulated. Indeed, as the case of one such richer migrant demonstrates, the information they carry often serves to determine the long-term trajectory of a migration:

> I came here to study English and Chinese at Pannasastra [university] where I paid $60 for a three month course. When I was there, though, I met a friend and fellow student who told me about working as a phone renovator. He told me I could learn this skill at the Bassac Market near Boeung Tumpun. I paid $100 and initially I didn't get any salary, but after three years I saved enough to open my own shop this year. (Veasna, 16/07/13)

Similarly, Sambath, now a *motodop* [motorcycle taxi driver], landlord and group leader [low level political administrator], related the following story:

> I first came to Phnom Penh to work as a *motodop* in the year 2000, but my old bike was constantly breaking down so I gave it up to become a labourer. Some people there told me about an opportunity in a factory, so I got my wife a job as a garment worker. With us both working, we saved money and I was able to buy a brand new motorbike for $1050 ... Now we own a rented room and make money that way, but we still work as a motodop and garment worker too.
>
> (Sambath, 01/07/13)

The need to gain access to the information which facilitates these opportunities means that ostensibly consumptive spending on leisure may be as important as money in terms of acquiring both occupational security and opportunity in the urban environment. However, as the qualitative data consistently highlight, the ability to engage in such activities is highly stratified. Only wealthier migrants reported an ability to attend 'small parties in rented rooms with drinks and food' (Borey, 21/05/13), 'visits to the market about once a month to buy clothes, spending about thirty dollars each time' (Savady, 25/06/13), or day trips to local leisure sites 'whenever we have a holiday' (Savady, 25/06/13). Poorer migrants, or more specifically those burdened by high remittance obligations, reported an opposite story, with all of those whose remittances exceed 50 per cent claiming to undertake no social activities at all. For these migrants, 'staying alone in [their] rented room' each evening (Vannak, 02/07/13) was ubiquitous.

This restriction on social network development is disadvantageous in various ways. Whether the target is to open a business, achieve promotion, or simply find less arduous work, information shared through incidental meetings underpins most stories of advancement in Phnom Penh. Participation in urban networks is therefore a vital means of improving either income or livelihoods more generally, but it cannot take place without capital. Urban networks, like any resource, cost money to maintain. Moreover, even once an opportunity has been identified, it must usually be seized at a price, whether in the form of a bribe to attain a given position, or the start-up cost of an independent business. A heavy remittance burden places both the socio-informational and economic ingredients of advancement out of reach.

3.4.2 Advancement

The drain of high remittance burdens on both financial and social resources means that migrants who send back greater amounts are less able to improve their urban circumstances. This is demonstrated in Figure 3.5, which divides

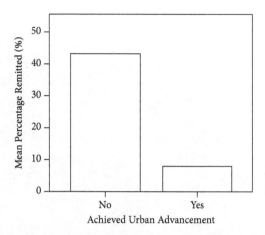

Figure 3.5 Annual household income as a factor in migrant eigenvector centrality (n = 50)

migrants into those who have significantly improved their circumstances and those who haven't. Almost all of those deemed to have 'advanced' had notably improved their income, but migrants' assessments of their own improvement have also been taken into account, so that progression into jobs deemed more presti-gious by their occupiers, such as from working in a shop to working in a microfinance institute in one case, may be deemed an advancement even if the salaries are comparable. This measure demonstrates a robust correlation with the percentage of income remitted to the sender household, as well as a strong correlation (Spearman's = −0.502, p = 0.01) with the rate (advancement/number of years in the city) at which this improvement was achieved.

From this and other data, then, a cohesive picture of rural–urban social relations begins to emerge whereby migrants from poorer households remit a higher proportion both of their own salary and their household's income. In turn, rural households appear to cultivate more central network relations within their village, potentially in order to mitigate the risk of their greater exposure to the modern sector. Migrants and their households may therefore be divided into those for whom the urban space is a long term opportunity, requiring the setting down of roots and substantial (economic and social) investment, and the many and varied others for whom the duration and type of urban labour is dependent to a much greater degree upon the varying requirements of rural household livelihoods.

For these latter migrants, modern sector labour often appears directionless and unending. Obliged by the needs of their household to 'work every day without stopping' (Pros, 28/06/13), such workers are vital to their family's livelihoods security, but envisage 'no future' (Pros, 28/06/13) for themselves. Unlike their

Figure 3.6. Workers austere housing in front of the luxury towers they are building in Phnom Penh, 2018. Inequalities at every scale are magnified in the city.
Source: Courtesy of Thomas Cristofoletti/Ruom/Blood Bricks/Royal Holloway.

wealthier counterparts, they have neither time nor money to integrate into the city, let alone take advantage of, or invest in the opportunities it offers. In this way, the inequalities of rural life are not mitigated, but magnified by the city, which offers opportunity and accumulation to its wealthier visitors, but only survival to the poorer (see Figure 3.6).

Conclusion

In the breakneck pace of contemporary Cambodian development, it is continuities and immobilities that stand out above all else. As recently as 2017, Cambodia was the joint 6th fastest growing country in the world according to the World Bank (World Bank Group, 2017). Many fortunes have been made and many more are in the process of accumulating. Yet focusing only on the growing distance between Cambodia's elites and the remainder of the population conceals a far more widespread process of social differentiation.

Indeed, as money flows through the countryside like never before—$500,000,000 per year from the garment sector alone, according to estimates by CARE International (2017)—rural areas have become increasingly marketized. A purely agricultural life is increasingly impossible. The money to farm has to come from somewhere, yet the manner in which this money is obtained reflects deep rural divides. In the face of steeply rising costs for essential farming inputs and wage labour, rural families are finding themselves unable to produce the crops that were formerly the basis of their livelihoods. They are left, in consequence, with a three way choice between debt, land sale, or migration away from the village.

By adopting a perspective on remittances that is both translocal and holistic in its analytical frame, this chapter has sought to demonstrate the value of an approach characterized by depth rather than breadth. Far from the egalitarian insinuations of Cambodia's 'rising tide', it has found a pernicious, lasting and mobile inequality. Moreover, it has highlighted the dual impact of capital flows: bi-directional economic and social remittances are not only an efficient means of distributing resources, minimizing risk, and ensuring the long-term survival of the family, but a means by which inequality is transferred from rural to urban areas.

From a migrant's perspective, therefore, financial inequality is determined both by income and outgoings, with the latter being more important to urban livelihoods in many cases. Differences in the scale of migrants' remittances dwarf not only other urban outgoings—such as the cost of rent or food—but even differences in urban migrant income. Remittance levels, which vary considerably between individuals, may therefore be seen to be one of the key factors determining the quality of urban migrant livelihoods.

Secondly, remittances are not a discrete financial flow, but are intimately tied to the functioning of urban and rural social networks. Migrants alter their behaviour according both to the needs and endowments of the household and their development of urban social networks reflects this, linking urban social integration to household wealth via the intermediary mechanisms of remittance burden and inter-occupational movement.

Remittances therefore engender not only current inequality amongst migrants but also future inequality, by restricting urban social network construction. The importance attributed to these networks within the qualitative data suggest that the differences in urban 'advancement' visible between differently remitting migrants is attributable to the inability of migrants with heavy remittance burdens to build inter- and intra-occupational networks. By contrast, migrants who remit little enjoy the benefits of additional capital and associational networks synergistically, thereby helping to replicate rural inequalities in the urban space.

From the growing army of factory workers to the dwindling population of smallholder farmers, then, Cambodians are ensconced in a 'net of obligations' (Meas, 1999) that directs their actions and structure their well-being. Rather than a story of aggregate growth, development here is therefore a contest played out between groups, families and networks.

4

Sowing and Sewing Inequality in the Home

The Everyday Experience of Translocality

It is something of an axiom that economic change leads to changing social norms (see e.g. Adger, 2000; Platteau, 2000; Putnam, 1995). New circumstances present new opportunities that do not fit with long-established practices, shifting traditional viewpoints and reordering the practices of those party to the changing economy. Nevertheless, whilst there is a good deal of truth to this perspective, it one that treats as an aggregate what in reality is subtle and intricately structured. Norms are not transmuted on a national or community scale basis, but as the result of individual negotiations, often within the household. It is from these small-scale contestations that new consensuses build and coagulate into discourse.

As a result, single narratives express little of the differentiation underway in the course of normative change. Conservatism may be either a luxury for the better off, or a necessity borne of poverty, whilst in times of need moral rectitude is judged through the prism of money. Each household builds its own new norms and the liberating power to provide may become a heavy burden if the demands of dependency are too great.

Moreover, as this chapter aims to show, the negotiations that reconfigure power within the household are fundamentally translocal in character. Never the bounded and unitary entity presumed by economists and planners in decades past (Brickell and Chant, 2010), the household is now more spatially and conceptually dispersed than it has ever been. Power is contested in neither rural nor urban areas, but between them, as harvests and fertilizers; salaries and leave, become intertwined in the calculus of morality and livelihoods alike.

This chapter will explore these new, translocal, power dynamics, by examining their linkages to the everyday logistics of mobile lives and livelihoods. In doing so, it draws predominantly on four periods of research: a large-scale quantitative and qualitative survey undertaken in Phnom Penh's migrant worker enclave of Teuk Thla and complemented by linked rural fieldwork in 2010; a further linked study of rural out-migration to the Setbo garment worker enclave in 2013; a linked, rural–urban qualitative study of elder migration undertaken in 2015; and a series of further qualitative interviews undertaken in 2017 in order to update and mutually integrate the findings of these studies.

What follows will explore the findings of this research in four parts, first outlining the changing nature of household livelihoods in the Kingdom, before

Going Nowhere Fast: Mobile Inequality in the Age of Translocality. Sabina Lawreniuk and Laurie Parsons,
Oxford University Press (2020). © Sabina Lawreniuk and Laurie Parsons.
DOI: 10.1093/oso/9780198859505.001.0001

linking these changing practices to specific normative changes in both gender and demography, as well as the reactions these shifts have engendered. Finally, it will highlight the distinct paths in evidence within this overall picture of change, using competing testimonies to demonstrate the divergent character of normative renegotiation in Cambodian households.

4.1 Change and Mobility in the Cambodian Home: Relational Translocalities

As elsewhere, the home in Cambodia has long been a more complex entity than tends to be appreciated. Commentators have long referred to the mobility of the Cambodian populace (Ebihara, 1968; Kaleb, 1968), yet entrenched assumptions over the static nature of gender roles and the unitary nature of the household (Brickell and Chant, 2010) have contributed to a power-blind conception of mobile livelihoods. Migrant remittances, for example, have been described as 'a driver of women's financial inclusion' (UNCDF, 2017: i), but whilst this is undeniably true in some cases, it is equally untrue in others. Far from promulgating coherent normative change, the opportunities afforded by the growth of Cambodia's modern sector have given rise to fierce contestation, as changing patterns of mobility have engendered new perspectives on the household and its occupants (see Figure 4.1).

The scale of Cambodia's mobility is essential to appreciate here. Over 4 million of Cambodia's 15 million population is an internal migrant, with an estimated 1 million more working abroad (UNESCO, 2018). Moreover, even those not recorded in these statistics are now enmeshed in systems of mobility, as a changing climate and growing number of land sales to companies and individuals alike (see Chapter 5 for more details on the various dimensions of land transfer) has rendered the rural economy increasingly dependent on flows of urban capital (Rahut and Micevska Scharf, 2012).

At the centre of this process has been changing approaches to rice-based agriculture. The staple continues to account for 68 per cent of caloric intake nationwide and some 30 per cent of total household expenditure (IFAD, 2014), making it a key concern for rural farmers and urban migrant workers alike. Yet high input and low output prices, combined with an increasingly unstable climate for smallholder rice production has engendered widespread changes in rural farming practices. In particular, as the growing irregularity of rainfall during the wet season necessitates ever more rapid reactions to changes in the weather, the slow, labour intensive methods involved in traditional rice growing have come to seem increasingly problematic to rice growers. Many farmers have correspondingly adapted their farming practices from the traditional transplanting method—wherein rice seeds mature initially in a 'nursery' plot before transplantation to a

Figure 4.1 Home in the village: a couple share chores outside their rural house in Prey Veng province, 2018.
Source: Courtesy of Thomas Cristofoletti/Ruom/Blood Bricks/Royal Holloway.

larger field—to the broadcasting method, in which seeds are scattered over the plot where they will mature.

This latter approach has the dual advantages of shorter maturation times—essential in an increasingly capricious environment—and lower labour requirements, needing only about 2 work-days/ha compared with 20 to 30 work-days/ha (Liese et al., 2014). However, it is considerably more expensive in terms of farming inputs, signalling a reorientation of the rural factors of production away from labour and towards more capital intensive farming methods. This has engendered significant changes in rural livelihoods. Agricultural wages have almost tripled since 2005 (IBRD and World Bank, 2015) in response not only to the departure of part of the labour force to the modern sector, but also the greatly diminished need for labour in the village economy. Villagers who once relied, wholly or in part, upon rural wage labour for their livelihoods have been forced to find alternatives, leaving the supply of labour available for planting and harvesting tasks both limited and often unwilling to undertake such work. Indeed, as one large landowner complained:

> I myself need a lot of workers to do a large farm. I pay $7 per day plus three meals per day, but still many people don't want to work for me because they are shy and ashamed to work for someone else in the same village, so they go elsewhere. It is very difficult to find enough workers here.   (Visal, Krang Youv, 28/05/2018)

As the testimony of this large landowner highlights, the shifts in mobility prom-ulgated by economic and ecological changes cannot be separated from their social context. Mobility is inherently political (Brady et al., 2015; Cresswell, 2010) and the shame of poverty manifests not only through assets, but also the form and function of migration and work. Consequently, workers who formerly depended on rural wage labour are now unwilling to undertake it in a much reduced format, rejecting the disempowerment associated with immobility by couching it in the language of culture. Mobility and norms are therefore dynamically linked, as well as contested. Just as laboring in the fields becomes a battleground between narratives of shameful submission and idle shirking, so too does migration outside the village bring with it positive and negative viewpoints. Returning successfully from Thailand or Phnom Penh brings praise for both an individual and their household, but it also brings cultural resistance to changes in household relation-ships, often expressed through gender norms. As the Cambodian Prime Minister Hun Sen outlined some years ago, for example:

> '...in Cambodian custom, daughters are not allowed to go far from their homes, because parents are afraid of their girls being raped, and when a girl is raped, her family may be dishonoured.'
>
> (Prime Minister Hun Sen, cited in Hill and Ly, 2004: 108)

Perspectives such as this highlight how the politics of mobility manifest not on an individual basis, but inter-relationally. Mobility radiates beyond itself, drawing actors and groups connected along a range of axes into its sphere of influence. Consequently, movement alters what contains it: 'from the microscale of the motorbike to the macroscale of migration' (Brickell, 2011: 451) ideals of mobility are being reassessed in relation to both gender and the home. Nevertheless, to view these orders of movement in isolation is to elide the complex dynamism of the factors that drive them. Normative reassessment does not occur within a discrete realm of norms, but via interaction with external circumstances, incentives, and constraints.

In Cambodia specifically, the normative changes associated with mobility reflect a relatively fluid approach to the household division of labour, wherein the 'necessity to finish farming activities on time seems to surpass the norm of division of labor' in times of need (Ogawa, 2004: 366). However, this mutability of norms is a phenomenon widely noted in rapidly developing Southeast Asia. A range of authors have engaged with normative change in the household and home (Blumtritt, 2013; Rogaly and Thieme, 2012; Brickell, 2012; Brickell and Datta, 2011; Truong and Gasper, 2008; Conradson and Mckay, 2007; Silvey, 2001) as a way to de-privilege economic flows in favour of 'the complex forms of subjectivity and feeling that emerge through geographical mobility' (Conradson and McKay, 2007: 167).

In doing so, they critique the 'battlefield of knowledge' (Truong and Gasper, 2008: 287) through which mobility is interpreted, examining, in particular, how subjective accounts may be used to deconstruct and elucidate the meaning of far larger scale flows. Migration is therefore viewed from this perspective not a process of dislocation, but of renegotiation and contestation, with the 'left behind' literature on non-waged members of migrant households, (e.g. Yeoh and Huang, 2014; Toyota et al., 2007; Yeoh et al., 2002), in particular, having demonstrated how norms of femininity (Devasahayam et al., 2004) and motherhood (Yeoh and Huang, 2010) have undergone profound changes in response to the shifting household-scale economy.

Above all, these studies have highlighted the heterogeneity of normative responses to mobility, a crucial development in a context where 'in the past a pervasive myth of family solidarity and unity tainted the approach of academics, development practitioners and policy-makers' (Brickell, 2011b: 1355). Understanding changes to the norms underpinning mobility therefore 'requires exploring the dynamics of broader forces conditioning them' (Resurreccion, Sajor, and Sophea, 2008: 17), including close attention to both ecological and economic factors. Without these, neither normative nor physical mobility may be satisfactorily explained.

Indeed, the vital and interlinked role of economy and ecology in driving mobility is of particular pertinence in contemporary Asia; 'a realm in which new realities are producing significant challenges for natural resource management, livelihoods and the mitigation of social inequalities' (Elmhirst and Resureccion, 2008: 3–4). To a degree, though, it applies everywhere. As scholars now recognize, not only are intra-household relations 'a continuous process of negotiations, contracts, renegotiations and exchanges between household members' (Brickell, 2011b: 1355), but these negotiations are themselves rooted in the systematic, complex, and multi-scalar distribution of resources. Rice, fertilizer, urban wages, and microfinance debts all figure in the household calculus of resources and roles. What is deemed to be acceptable, or even laudable, behavior depends on all of these specific endowments in combination. What follows shall outline how these changing patterns of resource use have altered relations along multiple axes of differentiation.

4.2 Stuck Inside of Mobile: Leavers and Their Ongoing Ties to the Home

4.2.1 Women, Earnings, and the Home

In the context of Cambodia's rapid recent development, it is important to bear in mind continuity as well as change. Despite the rapid growth since the mid-1990s of the garment industry, whose 90 per cent female workforce (CARE, 2017) now employs around 1,000,000 workers (ILO, 2018), 75 per cent of Cambodian women continue to work in agriculture (Layton and McPhail, 2013). Nevertheless, to view these women and their households as 'static', or merely the 'left behind' of migration would be to misrepresent the flexibility with which agricultural roles have been re-interpreted in recent years. Whether directly involved in migrant work or not, it is increasingly incumbent on women to adopt a portfolio of pastoral and income generating roles, many of which were previously ascribed to men (Elmhirst and Resurreccion, 2008). This has engendered greater mobility in some cases, a significant shift from even relatively recent times, when it was 'considerably limited' (Ogawa, 2004: 368). More broadly, the increasingly trans-local nature of the household has had a profound effect on the norms and attitudes surrounding women's work. Indeed, as a garment worker, from Kandal province, explained:

> Since I can make money now, if I say anything to my husband, then my husband has to listen and respect me. Before, everything I said was useless. Now if you make money, then when you talk to them they feel intimidated.
>
> (Female garment worker, Setbo, 16/06/2017)

These changes are felt throughout communities. Cambodia's economic development has had a profound effect on women's status in the home and as many rural Cambodians observe, 'now women have more value. They have their jobs, they can make money. They have more than before' (farmer, Krang Youv, 16/05/2017). Those able to make a substantial contribution to household income therefore report a much greater degree of authority over decision making and a louder voice within the household, yet changes of equal significance have also greeted those who do not command a salary. Most notably, these changes are reflected in widespread changes in women's mobility. Where previously 'more men than women went out of the village, and more frequently', engendering a 'distinct gap in the frequency of going out of the village between women and men' (Ogawa, 2004: 368), this distinction has narrowed or even reversed in some cases, as both men and women now migrate for modern sector work and the complementary mobility strategies that accompany it, including visitations and childcare. As a regular migrant grandmother explained:

> It's traditional for women to look after their children; it's happened like this for a long time. But back then they didn't need to go far from their homes [to work]. Back then, women worked at home and in the fields, [but] no one has any problem with factory women or [working] mothers today. Women's lives are better now: they have more knowledge, they can go to school, and even if they're illiterate then they can go to work in a factory. (Nuon, Phnom Penh, 28/07/2017)

As Nuon's testimony highlights, therefore, many of the changes in women's practices of mobility are enacted not as a result of shifting norms, but the maintenance of norms in changing circumstances. In particular, the attribution of childcare duties to women remains strongly in place in many cases, yet the manner in which the duty is enacted has changed. Women, such as Nuon, have had to significantly increase their mobility in order to continue to perform both their childcare and agricultural roles, criss-crossing the country on a regular basis so that urban salaried migrants may retain both access to their children and a basis in rural agriculture without needing to pay for costly childcare services in the city. Consequently, these alterations in the interpretation of norms do not occur en masse, but as part of a 'practical means-ends' renegotiation, rooted in the quotidian realities of translocal labour (Brickell, 2011: 453). Moreover, they may change on a seasonal basis, as the increasingly unpredictable fruits of rural agriculture rise or fall in value. For example:

> Now most women demand equal rights, the same as men. Now it seems like they have work in the factories and then the men have to work in the farm. It's not like before, where the women only work in the kitchen and stay at home. Before they had to transplant the rice, but since they have factory work, people go out. Nobody comes here to transplant, they just broadcast. If the man does the farm and gets a profit, then they stay in the village, but if they don't make any money from farming then they go out.   (Farmer, Krang Youv, 16/06/2017)

Such accounts highlight how women's mobility, as well as the norms of gender that underpin it, are not fixed, but subject to the caprices of environment and economy. In this way, 'gender itself is re-inscribed in and through practices, policies and responses associated with shifting environments and natural resource management' (Elmhirst and Resureccion, 2008: 9). Yet this occurs neither directly, nor in isolation. Normative change, though frequently rooted in the practicalities of livelihoods, is also relational; it involves resistance as well as adaptation and encompasses—in various respects—losers as well as gainers. Masculinities, in particular, have had to adapt to these new dynamics, ceding authority within the home in some cases, but seeking to re-inscribe traditional household hierarchies in others.

4.2.2 Masculinities and Relational Norms

Since the beginning of its mass expansion in the 1990s, the gendered interpretation of 'household' tasks such as needlework and indoor labour (Derks, 2008) have seen garment factory work taken up overwhelmingly by female workers. Nevertheless, what was once a matter of self-selection is increasingly becoming entrenched in hiring practices, as the growing influence of unionism within the industry—and in particular, the large scale, politicized protests that have accompanied minimum wage negotiations since 2013—has seen industry leaders seek out means to weaken or mitigate the influence of organized labour (Arnold, 2013). A key strategy in this respect has been to minimize the recruitment of men, who are seen as leaders and the instigators of disruption during protests (CARE, 2017). Consequently, even as the garment sector grows in importance within the national economy, factories have begun to actively favour female workers.

Given the relatively small size of the construction industry in Cambodia— between 50,000 and 200,000 employees (CARE, 2017; US State Department, 2015) compared to a million in the garment industry (ILO, 2018)—there exists a significant deficit in the demography of modern sector work. Working age men unable to enter the construction industry find it difficult to enter the garment industry, leading to a rise in rural underemployment, as land holdings concentrate and the demand for rural labour dwindles. As elsewhere in Asia (Elmhirst, 2007), this has led to delinquency and gangsterism (Czymoniewicz-Klippel, 2013), yet narratives surrounding the phenomenon suggest an alternative causality. As is often the case in narratives of unemployment (Howe, 1998), the absence of work is readily interpreted as distaste for it, with cause and effect becoming inextricably intertwined in relation to poverty and anti-social behaviour. A recognition of the poverty of the area does not therefore exonerate those with limited work from the character flaws traditionally associated with indolence, as a resident of one such village explained:

> These kind of people [gangsters] don't migrate because they need to be free and they don't want to live under other people's pressure. They are difficult people.
> (Mith, Krang Youv, 20/06/13)

Thus, from the perspective of narrative, just as poverty of rural resources is little defence against inactivity, nor either is the absence of work outside the village. Despite a lack of opportunity in many cases, men who are unable to find migrant work are viewed as lazy and unwilling to help their families, placing women in the role of gatekeepers to the garment industry in many cases; more easily accepted and in certain cases able to bring male family members into factory work with them. As one explained:

> The factories don't want the men because the men are trouble. But for my husband, since I saw him working at farming at the time I wanted him to come to work in the factory, I asked them to accept him
> (Female garment worker, 19/05/2017)

This dominance of the garment industry by women feeds into a wider narrative surrounding men and migrant work which contrasts unfavourably with the 'dutiful daughters' (Derks, 2008: 170) discourse commonly espoused in relation to women's migrant labour. Narratives of remittances, in particular, have become strongly gendered in recent years and a widespread discourse—borne out in part, though nuanced, by the statistical evidence (see CARE, 2017; Parsons, 2016)—holds that 'now it is better to have daughters than sons because daughters help the family' (Village Chief, Prey Veng, 04/04/2010). This perspective is often supported by female migrants, who note a difference in behaviour and spending habits between themselves and their male counterparts:

> We don't have money to go anywhere, because all the money is spent on our families. The men generally go to the restaurants round here, or small shops nearby... The men like drinking, so they go there. The women feel like they have more responsibility. They have children and family in the home village. Some men send more, some send less, but compared to the women, it's less than them... [Amongst men]...50% are good, 50% are not. I don't know [why] but people don't care about their families. They only care about themselves and just follow their friends.   (Female garment worker, Setbo, 16/06/2017)

This narrative has brought an enhanced role for women in some cases, but it has also brought strain. The demands of home and formal sector work are onerous and often incompatible, yet as translocality reconfigures the landscape of the household, women are increasingly bearing this double burden, leading isolated lives or endangering their health through inadequate nutrition (Enfants & Developpement, 2015; PSL, 2014). Moreover, even successfully balancing these duties presents difficulties, as male householders seek to differentiate temporary status gains from historic positionalities.

Most commonly, this is undertaken by narratively underscoring the primacy of tradition over pragmatism, with the dynamism of women's roles being set up against masculine norms styled as static, natural and rooted in tradition. Just as women are placed increasingly in the role of gatekeepers to the modern sector, therefore, 'boys and men are essentially "gatekeepers" to gender equality, often controlling women's access to the resources needed to claim justice in the domestic sphere' (Brickell, 2011b: 356). For married women, substantial role change is therefore only possible with the implied consent of a husband, who may resist the assignment of a less central role with reference to traditional hierarchy. As one farmer explained:

> Attitudes depend on the woman. Some women, [even] if they make money then they still respect their husbands, but other women, if they make money then they

decide [financial matters] themselves, because they want to get above their husband.   (Farmer, Krang Youv, 16/05/2017)

The result is that 'some families listen [if a woman isn't working], but some families don't' (female garment worker, Setbo, 16/06/2017), preferring instead to prioritize pre-existing gender hierarchies 'by presenting tradition as a static force... [and] ... household-level changes (such as the equal sharing of cooking or women spending more time outside the home) as a violation of men's rights' (Brickell, 2011: 448). There are two main features to this position: first, that 'traditional divisions of labour are temporarily negotiable only in infrequent circumstances and second, that it is men's duty to uphold to this custom, hence the idea of "dutiful non-participation"' (Brickell, 2011b: 1361). Both are aimed at severing or obscuring the relationship between economic and decision-making power within the household through reference to negative stereotypes of femininity, either within the household or in the context of modern sector work.

An example of such a narrative was expounded by the owner of an accommodation block in the garment worker community of Setbo (see Figure 4.2), who lamented that 'now the young girls always bring boys back to their room in the evening, who then leave in the early morning. Only about one in ten of them is good' (shopkeeper and rented room owner, 21/06/13). Moreover, denigration of this sort is not limited to the garment industry, but reflects a wider narrative of gender in relation to mobility:

Figure 4.2 Home in the city: a typical worker's rented room accommodation on the outskirts of Phnom Penh, 2018.
Source: Author.

I think that a woman can't live on her own because of tradition. A woman is supposed to stay at home. A woman living alone is like having a toilet in front of the house—it does not smell good to the other people.

(Male informant cited in Brickell, 2011: 452)

Variants on this narrative are repeated and referred to wherever domestic conflict arises. In particular, physical separation from the rural household, bemoaned as being 'not a good change because... women shouldn't be so far from their home' (Sar, Prey Veng, 29/07/2015), forms part of a wider narrative—familiar in other contexts (see Parsons and Lawreniuk, 2017b)—wherein women are criticized for their participation in a broader process of 'modernity'. This narrative and the critiques that accompany it predate the mass migration characteristic of Cambodia's contemporary economic environment, yet this broader discourse is informative in that it highlights how closely intertwined are narratives of mobility itself in the wider discourse of gender and the household. Stasis, viewed thus, is a highly political act, yet one whose discursive and physical manifestations differ markedly.

4.3 The 'Great Mobility' of Staying: Challenging Stasis in Mobility Narratives

4.3.1 The Role of Older People in the Context of Changing Agriculture

Even as the gender dimensions of the household have received growing attention in recent years, demographic concerns have drawn consideration only in a secondary sense; an orthogonal axis of inequality that intersects and influences, but ultimately does little to drive normative changes within the collective practice of life and work. In broad terms, this reflects a global research agenda in which older people 'remain noticeably under-researched in human geography' and beyond (Tarrant, 2010: 190). Yet this scholarly marginality masks the centrality of demography to the practice of livelihoods. Migration itself is most commonly undertaken by the young (Lim, 2007), but the decision to migrate is made in collaboration and contestation with 'non-migrants' selected by both age and gender. Stayers and leavers by the must therefore be considered with equal clarity, as outlined by the father of a female garment worker, now working in Phnom Penh:

10 years ago or more... I told my children that if they don't work then they won't have any money for their children, so they asked me if I could look after [my grandchildren] while they went to work. I agreed, but at that time it was very

hard work because it was two children under one year old. [Nevertheless] it wasn't really my daughter persuading me, but me persuading my daughter. I explained that if she loved her children then she must go to Phnom Penh and find work to feed them.   (Sarun, Prey Veng, 29/07/2015)

Sarun's account of his daughter's migration highlights not only the role of household norms in driving mobility, but also their dynamism in response to external circumstances. As with gender, the relation of household power relations to demography has shifted with Cambodia's economy to the extent that younger members of the family now wield much greater influence than they historically have done. Indeed, as informants outlined, it is those commanding a salary who wield the greatest influence in household affairs, even to the extent of superseding traditional age-based hierarchies. The spread of telecommunications in Cambodia (Phong and Solá, 2015) has catalysed this process, allowing migrants to control even rural household affairs whilst undertaking urban work. Consequently:

Generally, the older people just look after the children and do the farming. The children go out to work...Generally...before they do anything or make any decisions, they have to ask the children in Phnom Penh.

(Farmer, Krang Youv, 16/05/2017)

As with Cambodia's changing masculinities, however, status change is not a zero sum game. Rather, it entails consequences and knock-on effects in the sphere of those for whom status has been lost, as norms and practices change to reflect a variety of new roles. Migrant work does not simply leave a vacuum in sender areas, around which pre-migration practices remain unchanged. Rather, the logistical challenges that accompany such mobility enact complementary changes to the mobility of 'stayers', a phenomenon greatly under-examined despite its importance to the economics of migrant work.

So great is the need for supportive forms of mobility that many Cambodian grandparents today enact patterns of mobility far more rapid than their children, travelling, for instance, 'every Saturday morning and returning every Sunday afternoon in order to take care of my other children and bring them money' (Thorn, Phnom Penh, 21/07/2015). Whilst less rapid cycles are more common—fortnightly or monthly cycles visitation patterns to bring children are typical—the role of their mobility is crucial. Indeed, those migrants who are unable to work due to their lack of a relative to look after their children often bitterly bemoan the lack of a healthy elder relative to assist in their migration. As one explained:

My mother could never take care of my child because she was always sick and when the baby was one year old, she died...If you have [a grandmother to help raise the children] everything seems very easy. You can send the child to them and then both of you can work.   (Thida, Phnom Penh, 22/07/2015)

The lamentations of housebound women like Thida demonstrate the key role of those whose physical mobility does not extend beyond their home village in determining the outcomes of migration. In an environment in which 'the problem [of drought] has become very serious, with 'this year and last year, no rain' (Sar, Prey Veng, 29/07/2015), all members of the household—not simply those of the right age to enter the modern sector—have had to change their behaviour to adapt. Consequently, the newly arduous roles adopted by elder migrants were viewed not as an unwanted burden, but as a welcome means of retaining a useful role within the household. As an elderly man in a rural area explained:

> If we don't have grandchildren to care for, old people just stay at home. We have no other work to do. It may be easier to live, but the feeling is worse: if we don't look after children, then the children can't work and we feel bad about that.
>
> (Sarun, Prey Veng, 29/07/2015)

Furthermore, though rooted in specific household logistics, these normative changes are enacted in dynamic relation to both sending and receiving communities. Older people who undertake this work are viewed by their communities not only as avoiding marginality, but as actively participating in migrant work. Thus, pastoral and logistical support to migrant householders serves as a means to regain the status lost through exclusion from the modern sector. As a grandmother engaged in such work explained:

> People [like me] are respected for doing this. People say 'very good, she is helping', so it helps both of us [for my daughter] to work in the factory. [Doing this] is important now, because in my village only the men go out to work, so if the women can look after the children then it's ok, but if a grandmother can look after the children then both can go out to work, so compared to the people whose grandmothers don't look after the children it's much better.
>
> (Sopheap, Phnom Penh, 22/ 07/ 2015)

The inter-relational dimension of these developments means that changes to norms and practices amongst one demographic often depend upon changes enacted elsewhere. Just as women have not seen their social standing increase en masse, grandparents, similarly, do not lose or regain it via community reassessments in themselves. Rather, status, norms, and their interpretation are to a large extent dependent on idiosyncratic household circumstances: parental activities respond to that of their children and vice versa. Both gender and age, therefore, are closely intertwined, as outlined by one respondent:

[Traditionally] we don't have to look after the grandchildren for so long, just for a few hours whilst their parents are in the fields!...It's not traditional for mothers to work either [though]; only in this new society. It wasn't usual for women to go so far away from the village, but no one discriminates against them now because it's normal.   (Sovann, Prey Veng, 29/07/2015)

Accounts such as these paint a pragmatic picture of normative change that stands in contrast to several of those outlined above, where the changes to roles and mobility associated with migration have become a battleground of contestation. Here, the 'stayers' of migration, especially older people, appear content to abandon their traditional roles to adopt new patterns of mobility, whilst their children and peers are content, even grateful, to accept and affirm their dynamism. To some extent, this is indicative simply of the variance in community attitudes which might be expected across the complex landscape of Cambodia's development. However, as the next section will outline, there is a more structured dimension at work also. Who faces resistance and who flexibility is not merely incidental, but the product of a range of intersecting structural factors. Norms, otherwise put, are as unequal a resources as the structures that drive their adaptation.

4.3.2 The Multiple Trajectories of Inequality within the Home

Migration is increasingly becoming a key axis of distinction in Cambodia. Indeed, so ingrained has the importance of modern sector wages become within the structure of Cambodian society, that informants report a growing segregation between migrants and non-migrants, rooted in perceived wealth differentials, mobility, and shame. For example:

People get better off [after migrating]. They have money. So at the time that they meet [up again] they have a party. Also, when they come here and make money, they can call their relatives...Sometimes we call non-migrants to join, but sometimes they feel ashamed to join [us] because they think that when people leave the village they have money, unlike them. They don't have money, so they don't join [us].   (Factory worker, Phnom Penh, 19/05/2017)

In many ways, therefore, Hughes's (2001: 19) prediction that the Kingdom's development would lead it to becoming a nation of 'insiders and outsiders' has proved correct. However, the geography and scale of that distinction requires interrogation. Inequality of access to migration manifests not only between households and communities, but also at the far smaller scale of household relationships and family. Tensions, conflicts and reorientations shift within the home, shaping mobility and driving distinction; opportunity and obligation co-exist in complex interplay with norms of age and gender. Migration, in other words, is a phenomenon experienced through the lens of intimate relationships. Nevertheless, despite this inter-relational articulation, contextual factors play a key role in structuring normative change. As one smallholder farmer outlined, for example:

> [Whether we farm rice or not] depends on the children in Phnom Penh. Generally, if they make a lot of money then they will leave the farm, but if they aren't making so much, then they will come back to do it.
>
> (Farmer, Krang Youv, 16/05/2017)

That decisions of such cultural significance depend on transient urban factors is of particular note here. Livelihoods are a key determinant of how norms manifest and the changing livelihoods associated with Cambodia's recent development have seen normative values and practices shift significantly to accommodate them. Nevertheless, it is not norms themselves that are changing, but their interpretation. Practices have long been more fluid than tends to be recognized and divisions along the lines of, for instance, gender, are often rooted to a greater extent in practical considerations than normative fundamentals. As a result, whilst 'certainly there is a pattern of division of labor...gendered division of labor is not rigidly observed' but practically arrived at on the basis of a broad and re-interpretable typology (Ogawa, 2004: 365). As Ogawa continues:

> Generally 'heavy' work which requires physical strength is said to be men's work and 'light' work is considered to be women's work. Those requiring technical skills or involving possible danger, such as chemical fertilizer application and insecticide spraying, are regarded as men's work. Raising and selling small animals are mostly women's tasks as is selling rice. Plowing, insecticide spraying and ox cart driving are almost exclusively done by men and rice transplanting by women.

From this perspective, it is not only the environment that is viewed through a normative lens, but also norms that are viewed through an environmental one. The same norm may—depending on circumstance—produce multiple social arrangements, both within and beyond the household. Thus, 'the emphasis is on the ways in which changing environmental conditions bring into existence categories of social difference including gender' (Elmhirst and Resureccion, 2008: 9), as the interpretation of norms is guided by the environment.

Crucial to understanding this process is the structural dimension of ecology. As outlined by CARE (2017) and Parsons (2016) in the Cambodian context, as well as a range of authors elsewhere (Winsemius et al., 2015; Brouwer et al., 2007), the everyday impacts of ecology are articulated through socio-economic frameworks which channel the impacts of the natural environment towards those least able mitigate them. The norms that it engenders are therefore closely related to household resources. Migrants from the poorest households are most heavily burdened by the expectations of remittances and this is reflected clearly in the practice of livelihoods in both sending and receiving areas. Not only does the heavy impact of shocks amongst those with little to lose restrict the range of migrant choices, often resulting in poorer or less lucrative migrant work (Parsons, 2016), but often also the additional burden of debt accrued through migration undertaken in haste (Bylander, 2015).

As outlined in Chapter 2, therefore, the poorest migrants live isolated and deprived lives, sending back the majority of their salaries to their home household whilst retaining only a bare minimum for their own sustenance. The better off, by contrast, despite earning the same salary as their most penurious counterparts in many cases, enjoy a vastly superior livelihood as a result of their reduced need to remit. Such migrants are able to consume, socialize, and learn, building networks in urban areas that lead to progression into better conditions or more lucrative work in the future. What it means to be a migrant, in other words, depends more upon parental than personal endowments and rural ecology, in particular, casts a long shadow over migrant work.

As a result, filial and financial duties are increasingly intertwined. Though 'equivocal at best' in a general sense (Ahmed, 2004), linkages between economic contribution to, and voice within, the household are reported by Cambodian migrants in many cases (see Parsons et al., 2014). Of those migrants able to meet their financial commitments, several reported gaining a 'strong' role in their household, compared with 'before, when I was without work and not earning money [and] was not judged to be able to make good decisions' (Sonim, 05/05/ 2010). For those unable to meet these obligations, however, the result is criticism within the household. Children are accused of being 'lazy' and 'uncaring' (Sol, Krang Youv, 08/06/2013), regardless of the difficulties of securing modern sector work and the impediments to regular remittances even once successfully installed in it.

Those unable to migrate at all face still worse censure. As with their older counterparts, some younger women may succeed in retaining a voice within the household through supportive or pastoral activities in the home village, encouraging their families to 'listen to them because maybe they have children to look after and animals and chickens to look after' (female garment worker, Setbo, 16/ 06/2017), but the acceptability of this stasis depends on the mobility of others. Those not involved in either the supportive or wage earning dimensions of migration become the subject of widespread narrative denigration. As residents of one Krang Youv village explained, those who do not migrate—primarily men— 'just drink and play cards all day', or are idle and disinterested in farming (Khean,

28/05/13). Similarly, as a second underscored: 'these kinds of people don't migrate because they need to be free and they don't want to live under other people's pressure. They are difficult people' (Mith, 20/06/13).

These fragments of discourse highlight the crucial role of mobility in moulding the landscape of status both within and outside the household. The strength of a person's voice is linked closely to the perceived value of their role—often assessed largely in economic terms—but the availability of financially productive roles is limited and uneven, depending on the livelihoods and assets of each household. Mobility is key to gaining access to these opportunities, but is itself a scarce resource. The ability to access it is therefore not incidental, but linked into durable structures of inequality that play a significant role in determining how gender norms shift within a given household (Parsons, 2017). As Krang Youv residents explained, tradition and gender norms are bound up in practices of mobility which are themselves stratified by wealth:

> Generally the rich people have their own businesses in the village, so the women and men just stay at home. For the poor there are two ways: one is where the wife goes out to work and the husband stays to farm and look after the children. The other is that if they don't have any children, then they both go out to work. It's not traditional. Recently, since the factories came, the women have had to go a long way. (Farmer, Krang Youv, 16/06/2017)

In its allusion of the complex inter-linkage between norms, ecology and economy in Cambodia, this farmers' perspective highlights the cultural dynamism of migrant life. It demonstrates how structural inequalities determine economic livelihoods and the mobilities associated with them; how conceptions of tradition colour and influence these new mobilities; and the social bifurcations that result. Indeed, what is key here is not that the wealthy achieve respect within their household and community whilst the less wealthy do not, but that they do so in different ways. To be a dutiful daughter in this complex and diverse environment may be achieved via multiple trajectories: a decade's toil in a factory, or home-bound chastity in the village (see Figure 4.3). A father or grandfather, similarly, may gain praise either for the fruits of their economic labour, or their support of migrant women.

Yet these are not equivalencies. Each pathway is bound up in the complex inter-subjective milieu of what a person should and should not do, meaning that compliance with one normative structure may engender conflict with another. Shifting livelihoods therefore manifest not only in economic distinctions, but normative ones also, as communally bounded normative structures themselves shift and diverge. The result, as many poorer villagers have argued, is that 'the relationship between the rich and the poor is torn, as if it has been pulled apart' (Narith, 12/08/13), but to view normative change on this scale alone misses much.

Figure 4.3 Two generations of a family prepare food in a rural kitchen in Prey Veng province, 2018.
Source: Courtesy of Thomas Cristofoletti/Ruom/Blood Bricks/Royal Holloway.

Changes of equal magnitude are underway within households, where the meaning of each role has long since ceased to be a matter of community consensus.

Conclusion

The changes underway in Cambodia during the past two decades may be under-stood in a number of ways. Economic growth and ecological change have com-bined to catalyse a period of rapidly rising mobility, with hundreds of thousands now employed in migrant industries and households increasingly practising complex multi-local livelihoods. Patterns of mobility have shifted dramatically as a result, as economic feedbacks and inter-linkages reshape rural areas even for those who stay. Whatever the pattern of their movements—rural–rural, rural–urban, circular, pastoral, or even as member of the 'left behind'—all Cambodians are migrants now.

Understanding this process is key to understanding the Cambodian household for two reasons. First, it disavows the analyst of the false dichotomy of 'movers' and 'stayers' that continues to pervade much of the migration literature. Secondly, moreover, it begins to highlight how changing roles in one part of an economy, community, or nation may fundamentally alter the meaning of those elsewhere, both in spatial and scalar terms. As the character of farming livelihoods has altered, so too has being a farmer's daughter, wife, or mother changed in meaning. To be landless, static, male, female, young, or old, has in each case adopted a new tenor, with narratives of praise and denigration shifting in response.

As this chapter has aimed to highlight, such normative transfiguration is more than just the dependent variable in the wider story of economic and ecological change, but plays a crucial reflexive role also. In the complex, protean context of contemporary Cambodia, no two households are alike and individuals' burdens differ substantially in both weight and character to reflect this. This is not a nation that is adapting to modernity en masse, but one which is diverging at a variety of scales, not only between but also within households. As these livelihoods change and disperse, so too do the roles and statuses of those who pursue them. The home therefore not only becomes a site of internal contestation, but a focal point also for wider conflicts and divisions in the community, economy, and nation.

5

The Invisible Grabbing Hand

Translocal Ecologies of Economic Development

National ecology is a subjective resource; what is valuable is subject to consensus only rarely and encompassed by statute more rarely still. Consequently, in a country whose natural resources have been 'decimated' (Lawreniuk, 2017) by large-scale logging and land use change in recent years, big picture narratives have—perhaps inevitably—prevailed. Nevertheless, a closer focus is necessary. Whilst much of the vast upheaval of the Cambodian ecosystem has been underpinned by the unfettered provision of government land concessions since the late 1990s (Scurrah and Hirsch, 2015), the agglomerative effects of far smaller-scale economic processes may have instigated equally large-scale ecological shifts. Market forces, spatially embedded in the landscape via the mass labour migration that has transformed national livelihoods, have engendered land redistribution, land use change, and ecological degradation on an enormous scale. Yet 'Smith's myth' (Schleifer and Vishny, 1998: 10), the invisibility of the hand reshaping Cambodia's natural resources, has seen the translocal structures that bind these tiny processes of change pass largely unnoticed thus far.

In this regard, the 'hard and unbending doctrine' (Schleifer and Vishny, 1998: 10) of neoliberal economics has had a clear influence on the study of ecology (Le Billon and Springer, 2007). A subtle, but clear line of distinction has been drawn between government appropriated or allocated economic land concessions—often referred to as economic land grabbing—and the degradation of environmental resources as a result of corporate activity. For example, whilst the impact of Cambodia's voluminous labour migration is believed to be substantial (Diepart, 2015), detailed examinations of this impact are rare (Barney, 2012), with far greater attention directed towards 'the immediate impacts of the "enclosure" process associated with gaining access to land by investors' (Baird and Fox, 2015: 1).

A key issue, therefore, is stasis of perspective. The wider literature is invariably concerned to a greater extent with ecological changes engendered by unitary events than the interlinked mobile diasporas that instigate everyday ecological change. Similarly, academic investigations of the migration-climate change nexus have been drawn disproportionately towards the investigation of 'large scale events such as Hurricane Katrina in the United States or the tsunami that affected Indonesia, in which millions of people were displaced due to rapid and dramatic

Going Nowhere Fast: Mobile Inequality in the Age of Translocality. Sabina Lawreniuk and Laurie Parsons,
Oxford University Press (2020). © Sabina Lawreniuk and Laurie Parsons.
DOI: 10.1093/oso/9780198859505.001.0001

change' (Bremner and Hunter, 2014: 2), rather than the smaller scale, chronic pressures that dominate the everyday experience of climate change for most. Moreover, they have tended to focus on the relationship between climate and mobility in one direction only, exploring how the climate more or less autonomously instigates mobility, whilst rarely exploring how mobility may instigate climactic events and degradation.

Simply put, 'what is important, and which has not been studied systematically, is how land in one place (a result of both socioeconomic and biophysical agents) affects how people use land in other places (either nearby or far away)' (Baird and Fox, 2015: 13). This lacuna is the focus of this chapter, which adopts a translocal lens to bridge two key gaps in current understanding of ecological change. First, after considering the national literature in relation to the topic, it will examine how patterns of mobility link rural and urban land use change. Secondly, it will highlight the ecological feedbacks from urban to rural areas. Finally, the chapter will adopt these insights in a longitudinal context, to explore how land use change has been driven by the dynamically linked, rural–urban inequalities that characterize contemporary labour migration in Cambodia.

5.1 Ecology, Land Use and Mobility in Cambodia

The breadth and depth of Cambodia's land use change in recent years is remarkable, deservedly garnering considerable attention in recent years. In total, it has been stated, 'more than 2.2 million hectares of land have been leased out to both domestic and foreign investors under ELCs, equivalent to more than a half of Cambodia's total arable land' (Marks et al., 2015: 33). Yet as impressive as they are, even these figure under-represent the true extent of the changes underway in Cambodia. Outside of, but interlaced with, government concessions, the dynamic yet persistent linkages between migrant occupations and rural livelihoods (Lawreniuk and Parsons, 2017; Parsons, 2016) have seen an increasingly mobile populace engender vast patterns of land re-allocation and re-use. Nevertheless, despite a widespread awareness of the key role of migration and mobility in Cambodia's development (Diepart, 2015; Baird and Fox, 2015), these everyday patterns of movement continue to be marginalized in the literature exploring the wider implications of land use change and re-allocation.

Where such issues are explored, indeed, it tends solely to be in terms of the livelihoods that emerge after land use change, rather than as dynamic factors in that change and its effects. Thus, 'while new research on the cultural dynamics of the rural migration process has highlighted a number of critical issues, insufficient attention has been paid to the study of agrarian relations in sending communities' (Barney, 2012: 76). In particular, current conceptualizations linking land reallocation and environmental degradation in Cambodia have tended to root their analyses both in spatially bounded localities and the production of a particular resource. Studies in this vein have most commonly focused on the ecological impact of land reallocation linked to deforestation (Davis et al., 2015; Milne, 2013; Borras and Franco, 2013), agribusiness (Sherchan, 2015; Schneider, 2011) and the extractive industries (Global Witness, 2010; 2009), the scale of which, as these studies elucidate, is enormous:

> Over 770,000 people (almost 6% of the total population) have been negatively affected by land grabs and more than 2.2 million hectares of land have been transferred from villagers into the hands of the corporations and rich people under the economic land concession scheme.   (Marks et al., 2015: 42)

Similar issues are addressed by the Cambodian political ecology literature, which has adopted both a broad lens—exploring the Kingdom's recent development as a whole (Milne and Mahanty, 2015; LICADHO, 2009)—and a suite of more specific analyses. Amongst these, the logging sector (Milne, 2015; Singh, 2014; Le Billon and Springer, 2007; Le Billon, 2000) has received perhaps the most attention in recent years, but notable studies have also emerged concerning the impact of economic land concessions (Neef et al., 2013), community level payments for environmental services, Milne and Adams (2012), forest carbon (Milne, 2012), and Chinese dams (Siciliano et al., 2016).

Studies of this sort have proved invaluable explorations of the geopolitical implications of land use change and ecological degradation in Cambodia. Nevertheless, despite their thematic breadth, investigations of land use are characterized by static and locally bounded approaches (Scurrah and Hirsch, 2015; Baird and Fox, 2015; Barney, 2012), which treat the multi-scalar economic processes engendering mobility separately to the large scale population displacements occurring under the remit of state sanctioned land repurposing. Smaller scale changes in community level land use are therefore rarely linked back to the same geopolitical processes as their state allocated counterparts, despite occurring 'on a large enough scale to endanger not only the local environment, but also the climatic system as a whole' (Vigil, 2016: 15).

Consequently, 'these relationships are poorly conceptualized, lack systematic investigation, and are reduced to simplistic causal explanations' (Warner et al., 2009: 690). Yet they are vitally important. Whether engendered by corporate or 'endogenous' (Hunsberger, 2015) forces, changes in land distribution generate migration to the garment industry, construction industry, rural wage labour, and private transport operations (Marks et al., 2015), all of which leave their own footprints on ecology and land use.

Seeking to bridge this lacuna, researchers have proposed the need for 'an approach that looks at biophysical and socioeconomic impacts of change at multiple sites' (Baird and Fox, 2015: 13) in order 'to study intersections and complex interactions within and across social, ecological and institutional domains' (Hunsberger et al., 2015: 3). However, despite recent efforts (e.g. Mayfroidt et al., 2013) to incorporate linked, or 'telecoupled' ecological zones 'gaining some traction' (Baird and Fox, 2015: 4) in the former case, accounts of both ecology and migration continue to be characterized by static conceptual approaches to ecological processes. Consequently, 'even when stasis is not assumed, [migration] is still not given the same level of attention as other historical processes' (Alexiades, 2009: xiii–xiv).

Guided by these calls, then, this chapter brings together sources from multiple projects, to produce a 'multi-sited ethnography' (Marcus, 1995: 95) based in three locations: Phnom Penh, Krang Youv, a rural commune in Kandal province; and Setbo, a peri-urban area to the south of Phnom Penh, which is dominated by the garment industry and home to several thousand migrant workers. Interviews in these locations were conducted over a period of six months, between March and September 2013, according to a mixed method approach incorporating quantitative interviews, qualitative interviews and focus groups. The data presented in this chapter represents the results of 77 rural interviews, 50 urban interviews and 2 focus groups, conducted in 2013 and 2017. The quantitative data presented in Figure 5.4 were obtained via a recall methodology in which land transfers were pegged to key historical events in order to improve accuracy. All interviews were undertaken by a two person team including one native and one proficient Khmer speaker.

5.2 The Translocal Ecology of Labour Migration: Factory Work and Multi-sited Resource Degradation

Far from the bounded and static manner in which it is often portrayed, rural land use change in Cambodia is a complex, mobile and multi-sited phenomenon shaped by multiple inequalities. As outlined in Chapter 3, distributions of land impact not only on livelihoods decisions, but how those decisions manifest in the future. Equal or similar migrant incomes may therefore lead to vastly different outcomes depending on a household's rural circumstances. Moreover, the obverse too is true. Rural circumstances shape and direct the impact of migrant remittances in rural areas, entrenching inequalities and dynamically stratifying livelihoods at both sides of the rural–urban continuum. Under these circumstances, both the impact of ecology on livelihoods and the impact of livelihoods on ecology are fundamentally translocal phenomena.

Nevertheless, the ongoing tendency in the climate migration literature towards unilinear, single-factor models has tended to impede recognition of the 'multi-causal relationship between environmental, political, economic, social, and cultural dimensions' of climate mobility (Piguet, 2010: 517). Consequently, certain

aspects of the relationship are invariably elided or ignored in the national literature (Tong and Sry, 2011; Dun, 2011; Heinonen, 2006), where 'migration-
environment linkages are predominantly perceived as the negative effects of
migration on the environment' (Heinonen, 2006: 456). What such an approach
tends to miss is the systematic dimension of the migration–climate nexus:
migrants do not simply arrive in a place and degrade resources, but build
mobile—and to an increasing degree translocal—economic systems in which
environmental resources are incorporated. Ecological change, thus, is not engendered by migrants, but through them, as patterns of mobility link discrete agro-
ecological zones.

Despite their prevalence and importance, though, establishing the form and
extent of these linkages requires a broad perspective on migrant industries,
inclusive of both macro- and microscopic changes to livelihoods in multiple
areas interconnected by population flows. Conditions in this regard are not
ubiquitously clear, but are exemplified perhaps best by the migration systems
surrounding peri-urban areas, of which Kandal's Setbo garment community
provides an especially instructive context. Located some fifteen kilometres from
the outskirts of Phnom Penh, this site is a collection of formerly small riverside
villages, which has expanded in the course of the past sixteen years to become one
of the largest centres of garment work in Cambodia (CCHR, 2013).

The continued expansion of factories and the necessity of accommodation for
their workers has attracted a significant volume of construction workers, as well as
schools and medical centres, and a bustling roadside market economy selling
foodstuffs, consumer durables and various other services. As a local rented room
owner explained, since the first factory opened in 1997, the area has changed
almost beyond recognition:

> Since then, the houses around here have changed from palm leaves to concrete,
> and the young people in the area now have something to do. Local people get
> money to improve their houses by selling things to the workers in the factory
> and people also sometimes sell their rice land to the companies because you
> can get a good price... You can make a lot of money selling land: a hectare
> goes for [$]10k or 20k. It's very expensive. [As a result], people don't farm rice
> much anymore, just vegetables and *chamkar* [non-rice agricultural plots which
> require little land]. (Rented room owner , 21/06/13)

This personal account of the physical changes to the local environment is borne
out by satellite imagery, which, as demonstrated in Figure 5.1 shows a significant
amount of development in only four years between the earliest and latest available
images. In particular, the number of tin- and tile-roofed houses, which show up
more brightly in satellite imagery than those constructed with wood or other
traditional materials, appears to have expanded markedly, as has the number of

Figure 5.1 Setbo in 2006 (left) and in 2010 (right).
Source: Google Earth.

rented rooms, which show up as longer, tin- [i.e. blue] roofed buildings arranged tangentially to the road.

This transition has not been without cost. As garment workers reported, 'people's health here is bad. Generally people are exhausted and they have diseases. This is because the environment is not good. Sometimes the smells from the cotton give people headaches' (garment worker, Setbo, 16/06/2017). Moreover, in environmental terms, the satellite data present a clear picture of change. Not only has the cluster of factories to the left of the river expanded to incorporate new buildings, but so too has the area of chemically infused liquid waste and rainwater overspill, now a highly visible feature of the landscape from above. Though invisible from the main road, this encroachment has not bypassed the livelihoods of local people, many of whose farm land is flooded on a regular basis by liquid emissions from the factory. As a local shopkeeper explained, though, those affected have no recourse to authority:

> I have a problem with the factory. My land is lower than the factory, which has a flood pipe coming out of the side of the building which floods my land completely in rainy season. I have asked the commune chief to intervene, but whilst they acknowledge there's a problem, they don't do anything. This has been going on for nearly twenty years, since the factory opened. As a result, I never farm in the wet season, just a very small vegetable crop in the dry season.
>
> (Shopkeeper, 21/06/13)

Four years later, residents of the area reported similar issues on an even larger scale, complaining that pollutants from the factories were regularly being pumped into the adjacent river, threatening the fisheries that many local people depend upon for their livelihoods. As a local shopkeeper related: 'Last year we had very polluted water, very dark ... We complained to the provincial authorities because there was a factory that put fluid into the river and poisoned the fish, which died. But we still don't have any solution to this problem' (shopkeeper 16/06/2017). Moreover, such issues were not limited only to fishers. Local crops are also threatened by overflows of chemicals from the reservoirs of polluted water produced by local factories:

> Sometimes, if it is raining heavily, then the waste water goes out and spreads into the farm. When this happens, it kills the plants....Especially in the rainy season, the water is very polluted. Generally, we get a yield only from the upper land. If the lower land is flooded by the water, then it is already all gone.
>
> (Farmer, 16/06/2017)

As these accounts demonstrate, the direct ecological impact of translocal industries is often severe and may lead, through the lens of livelihoods, to changes in the land use of the area. However, indirect, economic factors are also key to land reallocation. The local economic boom fuelled by salaried garment workers, as well as the construction workers who build the factories which pay them, has led to marked benefits for those local community members who previously owned land, but far smaller gains for those who possessed little or none. Many of those who previously relied on rice and fishing based livelihoods have taken advantage of the rapidly rising value of their land, to invest in alternative, high value crops, or commercial premises the better to further exploit the growing opportunity in the area. One such beneficiary is the same shopkeeper whose land is regularly flooded by the factories, who went on to explain that:

> I previously planted vegetables, but now I've stopped because I am too old. Previously it was plantation land, but now all my children live on it. Since the factories opened, they've build rented rooms. People still do [farming], but only older people, as the younger ones work in the factory.
>
> (Shopkeeper, 16/06/2017)

As a second landowner elaborated:

> [Migrants] amount to about ninety percent of my business, because local people generally just do *chamkar* [vegetable plantations]. I also have rented rooms and regularly borrow from the bank to expand them...all of [my children] are vendors along this street.   (Shopkeeper and rented room owner, 21/06/13)

By contrast, those who were landless or near landless prior to the influx of the garment industry have transitioned towards livelihoods in an alternative economy. Numerous residents whose landholdings are, or were, insufficient to profitably farm *chamkar* now support themselves by trading in low value items such as hair clips and nail varnish, which are cheap to purchase and can be sold in high volumes for a small profit during workers' lunch breaks, or before and after shifts. Nevertheless, involvement in non-farm activities of this sort does not denote dislocation from the natural environment. Rather, the predictability and brevity of these vending periods during the day has led a number of the less well-off residents of the area to split their days between traditional activities such as fishing and agricultural wage labour, and daily periods of vending to migrants. As one such informant explained:

> In the morning, we go fishing in the river, but in the afternoon I rent a space on the floor outside the factory gates for 3–4k for two hours. I move from factory to factory, all along the road, as well as several in Setbo and Ta Khmao. I work with my wife every day, making a turnover of 50k and a profit of 10k. In the evening, I go to another factory, making another 10k. Around the time garment workers get paid, I earn a lot more. The most popular product, in general, is nail varnish, especially the red one.   (Make-up and accessories vendor, 08/07/13)

As such, an influx of migrant labour such as that experienced by Setbo has the potential to generate not only direct changes to land use and ecology, but also indirect changes through its impact on livelihoods and the local economy. Furthermore, translocal ecological linkages such as these are not unilinear, but form part of complex networks of knock on effects and feedbacks to other areas, as the case of a local informant—whose household livelihoods are split between migrant and non-migrant activities—exemplifies:

> One of my children is a garment worker in Setbo, another is a vendor in Toul Kork market, in Phnom Penh, another two work as *chamkar* farmers on my land, but also sell things in the market. My friends' children also work in the factories.
> (Rented room owner, 21/06/13)

As this case demonstrates, the changing ecological profile of Krang Youv has generated a portfolio of outcomes across multiple ecological zones: participation in the local modern sector market economy, land use change from rice to cash cropping in the surrounding areas, and migration from the peri-urban area of Setbo to Phnom Penh. Changes wrought to sender ecologies by reduction in the labour supply (IBRD and World Bank, 2015) and the influx of remitted income have therefore had a major impact on land use, as the availability of capital to invest in new techniques has placed growing pressure on scarce resources. Moreover, translocally linked ecological changes such as these operate multi-directionally, with feedbacks often proving to be as significant as, or more so than, initial conditions.

Migration, thus, is shown here to be a phenomenon which not only links ecological changes occurring in one places to ecological changes taking place in another, but one with the capacity to catalyse and engender shifts in translocally linked ecologies. Crucially, moreover, these 'telecoupled' alterations need not be of the same type or order, but may manifest as land use change in one location and environmental degradation in another. However, what has yet to be explored is the dynamism of these linkages in relation to broader social factors; how, otherwise put, the inequalities which persist across translocal systems manifest in structured ecologies separated by space but linked by durable structures of advantage and disadvantage. By tracing the migration patterns of Setbo's migrant workers back to their point of rural origin, what follows shall attempt to elucidate these linkages.

5.3   Translocal Ecologies of Land Distribution: The Impact of Modern Sector Wages

5.3.1   Electric Pressures and Climate Shocks: The Translocal Ecologies of the New Rural Poor

Located less than an hour from Setbo, the commune of Krang Youv is a key source of migrants both to the peri-urban areas outlined above and the capital, Phnom Penh. As elsewhere in Cambodia, it has proved prone to climactic shocks and environmental risk, the most severe of which was the nationwide flood (and in some cases subsequent drought) of 1993. After this, the years 2011 and 2012, wherein widespread drought and flood respectively destroyed a significant proportion of the rice harvest and reduced yields elsewhere, are recalled as having been notably problematic from an agricultural perspective.

Such shocks are invariably viewed in terms of their impact on the paddy rice crop. As the head of the commune explained: '94% of people here are farmers' (Commune Chief, 15/08/13), whilst nationwide survey figures estimate two thirds of the workforce to be employed in agriculture (FAO, 2014). More broadly, the image of Cambodia as a land characterized by unending paddy fields and villages of farmers (see Figure 5.2) to work them is a much loved one in Khmer culture and continues to dominates national—especially rural—discourse (Chandler, 2009). Nevertheless, if the prevailing wisdom ever told the whole story, population and economic growth, with their associated pushes and pulls away from a unitary household income, have played their part in dislocating this axiom from reality. Thus, in contemporary Cambodia the following assessment by a local large scale

Figure 5.2 A farmer surveys his rice fields in Prey Veng, 2018.
Source: Courtesy of Thomas Cristofoletti/Ruom/Blood Bricks/Royal Holloway.

farmer and businessman, represents a more balanced picture than that provided by the commune chief. As he explained:

> In Krang Youv commune, there are sixteen villages, divided into four compass points. In the West part, most people depend on fishing. Here in the south, they depend on rice… Ampil is dominated by the market. Everybody is selling and trading things. The fishing part of the village is very poor. They rely on fishing and the rest of the time they just play cards.   (Phirun, 28/05/13)

This neatly summarized distinction constitutes a useful approximate typology of Krang Youv. In reality, though, these groupings are neither static, nor mutually exclusive. A wide range of alternatives and complements to rice farming shape the rural economy (see Rigg, 2012 for comparative examples in Thailand) and modern sector incomes are key to investing in and accessing these. Moreover, rice farming is itself increasingly underpinned and shaped by extra village income sources. As discussed in Chapter 6, Cambodia's ongoing transition to alternative forms of farming (Liese et al., 2014) and uptake of new rice varieties capable of maturing over a shorter time span (Keo, 2014) are predicated not only on the need to adapt to a changing environment, but also possession of the capital to do so, and the cyclical feedbacks these two factors engender.

Simply put, 'many studies indicate that changing crop varieties and water management are the most common adaptive strategies being practiced by households in Cambodia' (Chhinh and Millington, 2014: 178), but acquiring the capital to do so requires a livelihood inclusive of modern sector earnings. Consequently, farmers who would previously have sought to retain their household labour resources locally in order to provide the extra capacity needed for labour intensive transplanting practices, are increasingly minded to 'allow their family members to migrate to find a job in order to earn money for supporting daily livelihood and for starting up agriculture production again next year' (Keo, 2014). Indeed, as the Commune Chief (15/08/13) further explained:

> People first started to migrate when they started to get two-wheeled [hand] tractors because it allowed them to replace their labour. It meant they didn't need people to help with transplanting and planting and harvesting.

Nevertheless, the capitalisation of rural areas has been uneven in its impacts. The poorest and most marginal members of rural communities, including those without the social linkages to migrate (Parsons, 2016), the means to secure a loan (Bylander, 2014), or those whose physical impairments affect their ability to work (Gartrell, 2010), are often unable to secure a modern sector income. Under such circumstances, they must therefore continue to pursue their livelihoods in an environment which is not only subject to growing climate pressures, but is also driven increasingly by external flows of capital to which they have no access.

In this respect, one of the most common means by which villagers have sought to circumvent a lack of access to modern sector income has been to take advantage of the growing accessibility of the booming Cambodian microfinance sector (Bylander and Hamilton, 2015). However, as critics of microcredit have elsewhere asserted both globally (e.g. Bateman and Chang, 2012) and in Cambodia itself (Bateman et al., 2019; Ovensen and Trankell, 2014) the benefits of credit under such circumstances may be equivocal at best. Microfinance is increasingly a fig leaf to the declining margins of smallholder agriculture, trapping struggling farmers in cycles of crop failure, migration and debt from which they are unable to escape, as one Krang Youv resident explained:

> If you're rich then you use your own money to buy [farming inputs], but if you're not then you borrow money for the fertiliser, etc. Now for all inputs, like fertiliser, the price is high. But when they sell, the price is low, so [farmers] always lose. This pulls them into debt. First they sell the rice to pay for all the fertiliser and petrol bills, then there's not enough left, so they borrow money from a microfinance institution. Then they migrate to work for the microfinance loans and sometimes even sell their land.   (Fisher, Krang Youv, 16/06/2017)

Moreover, as a farmer confirmed first hand:

> Now when I'm doing farming, I lose every year. If I lose on the farm, I borrow money from an MFI. If I borrow a little, it's OK because I can get a job in the village to pay it, but if people lose a lot [of money after a bad harvest], then they have to borrow a lot and then they have to migrate and not come back until the loan is paid off.   (Farmer, Krang Youv, 16/06/2017)

Indeed, it is a common pronouncement in Krang Youv that 'many people borrow from banks and microfinance organisations, but the people never escape their debt, even if they go abroad to try to earn something to send back' (focus group, 09/05/13). Even more damningly, the head of Ampil village sought to highlight the long term danger of easy credit availability for those with the least assets. As she stated, 'previously [the poor] did not borrow from banks because there were no banks to borrow from. According to government policy, such banks should be helping to reduce poverty, but in fact they are making people poorer and poorer. They eat today, but they get poorer and poorer in the future' (Village Chief, 29/04/13).

This cycle of deepening poverty for those with the fewest assets is driven in no small part by the need to invest in increasingly expensive farming inputs despite the outputs of agriculture being so uncertain that even the largest farmers—who tend to posses the most diverse and thus lowest risk land portfolios—to admit that 'with rice farming, sometimes you win and sometimes you lose these days' (Phirun, 28/05/13). Furthermore, it is not only the cost of inputs that it rising, but the volume required, as extensive chemical fertilizing starves the land of nutrients. Consequently, the self-sustainability of rural agriculture in previous years has been replaced by the compulsion to use chemicals. As farmers explained: 'before, I could just use cow dung and still get some yield, but now if I use no fertiliser then I get no yield. Now everything is chemicals' (farmer, Krang Youv, 16/06/2017). As a second rural resident elaborated:

> Farming is more difficult. It's not easy. Now, because the soil is less fertile, we can get only a low yield because of this. I don't know [the reason], but based on what I do, I previously used five or six bags of fertiliser and now I use more than ten, but the yield is still lower than before. I have to do it because I don't know anything else to do, so I just borrow money to do it. Farming now makes no profit, but I don't know anything else.   (Farmer, Krang Youv, 16/05/2017)

Especially sententious here is this farmer's awareness of the limitations on his options. Rice farming is not the only rural livelihood that is threatened by the twin forces of climate and the market and moving away from agriculture presents no escape from the cycle of debt and insecurity exhibited therein. Rather, inequality is deepening with equal rapidity within Krang Youv's fisheries, a sector upon which a growing number of people are being induced by their agricultural travails to rely, despite its being 'highly vulnerable to the effects of climate change' (The WorldFish Centre, 2009: 2). Indeed, although this sector provides the 'livelihoods for millions and up to 80% of all animal protein in the diet', it is being adversely impacted by changes in the viability of smallholder agriculture, as growing numbers of landless former farmers compete for fish stocks which are declining due to the effects of climate change (Marshke et al., 2014). Thus, as Marshke et al. (2014: 2500) explain:

> this is not an easy moment to be a small-scale fisher or to live in a tropical coastal area, whereby over fishing is felt and the impacts of climate change and variation are also starting to be felt. Fishing these days demands increased effort, and people question how long this might be sustained. At the same time, not everyone has the option to switch into non-fishing activities.

In this respect, Krang Youv is a microcosm of the national situation. Fishers report that 'the water is polluted by the chemicals and fertiliser that they use for farming' to the extent that 'it's poisonous for the fish ... [which] ... die because of it' (Bunroeun, Krang Youv, 16/06/2017). Moreover, as two further fishers independently reported, 'the quantity of fish is getting lower and lower ... and also it is getting hotter and hotter, making the water levels low' (Chhay , 22/04/13). This has already resulted in 'there [being] fewer and fewer fish to catch' (Vibol, 08/04/13), but as a neighbouring fisherman noted, the key issues faced by those involved in local fisheries have begun to compound each other, putting increasing pressure on fish stocks and forcing many to 'use illegal equipment which kills all the fish' (Rithisak , 16/06/13).

The 'illegal equipment' in question is the *masin chut*, a fishing rod modified to channel an electric current from a car battery to the surface of the water on contact. Although a crude instrument, it is notably effective, easily generating multiples of the yields obtainable from a day's fishing in the 'traditional' manner (see Figure 5.3). Certain users reported catching double what they could without the device and testimonies from elsewhere indicate even this could be a notable underestimate. Studies have suggested that up to five times the normal yield may be obtainable in this way (Kurien, 2007), figures which have inevitably rendered the practice illicit in light of the heavy and unsustainable toll it takes on water-borne environments.

Nevertheless, given that those most reliant upon fishing are amongst Krang Youv's poorest residents, few of the *masin chut's* users possess the ability to pursue alternatives. Consequently, in their rapidly marketizing environment, Ampil's fishers must either pursue declining fisheries without electricity, or the short term boon of an unsustainable but profitable income with them. Indeed, as a

Figure 5.3 A fisherman casts his net in Kandal province, 2018.
Source: Courtesy of Thomas Cristofoletti/Ruom/Blood Bricks/Royal Holloway.

villager who makes a living by recharging the batteries used by electric fishers explained:

> These batteries are for electrocuting mice, which are a major pest in the fields, but people also use them for fishing. Most of the people who use them to catch fish are poor people. The authorities have banned this [practice], but they continue because if they don't do it they will have nothing to eat.   (Khemera, 14/06/13)

Otherwise put, fishers in Krang Youv are faced with a choice between degrading their personal, or household, livelihoods, through the accumulation of debt, or degrading their communal livelihoods through the use of the *masin chut*. As those using the device well know, this sort of electric fishing within water resources already under severe pressure from the effects of overpopulation, land concessions and development, and changing rainfall patterns (Bansok et al. 2011) contributes towards a vicious cycle in which fishing with traditional methods is increasingly non-viable, thereby forcing more and more people into unsustainable fishing practices. The results of this tradeoff are already being seen across the country, as fishers nationwide compare better days 'two or three decades ago [when] fish were plentiful, especially big fish' to a contemporary situation in which 'fish catches are changing; fish are smaller and the size of individual catches has been decreasing' (Bansok et al. 2011: 33).

The process underway in Krang Youv is therefore one of rising pressures, as the merits of 'short-term *versus* longer-term needs of fishers and of the environment' are weighed against each other (Marshke et al., 2014: 2500). Moreover, the specifics of this conundrum are merely a microcosm of a wider trade off being undertaken by the landless or near landless across Cambodia, who must balance the needs of short-term consumption against the longer-term prospects of their household, often resulting in decisions which alleviate short-term pressures by entrenching long-term inequality. As the following section will show, these changes to the wider rural economy are increasingly rendered permanent as the influx of remittances and credit changes environmental risk 'from a socio-ecological to a socio-economic phenomenon' (Parsons, 2017: 147).

5.3.2 Remittances, Credit and Land Transfer: Translocality and the New Rural Rich

In stark contrast to the difficulties faced by the worse off members of translocal communities, migration for Krang Youv's wealthier inhabitants constitutes an invaluable catalyst to agricultural investment. Not only does it provide funds for investment in additional farming inputs such as better seeds and greater mechanisation, but also the rental of additional land from those unable, or unwilling, to arrange the necessary labour and machinery to work the fields themselves. As one such rental-farmer explained:

> I have two plots of my own land, 0.5 hectares of low and 0.3 hectares of high land. Now though, I also hire 15 hectares of land to do rice farming. We have been doing it for three years already and plan to continue... We do it twice a year and it costs about $8–9000 for fertiliser, gasoline and weed killers. We rent it from a neighbour who has a lot of land: nearly 100 hectares in one place.
>
> (Vuthy, 20/05/13)

As Vuthy's account demonstrates, the investment cost of hiring land on this scale may be sizable, available only to those with a significant amount of rice land to use as collateral, but the returns are even greater. For his family's investment and labours, Vuthy has thus far been rewarded with 1800 75kg sacks of rice each year in which he has undertaken this endeavour. Sold in this volume, he is forced to accept a discounted price from the usual 1000 riel [0.25 cents] per kilogram, but even at 750 riel [19 cents] per kilo his annual income from farming exceeds $25,000, leaving a profit margin of over $16,000.

Given that the gross income from his own farm land amounts to less than $1000, the profit available to those who invest in this way dwarfs almost any other available avenue. Income from fishing (including fish used for household consumption) generates an average of $1592 per household per year and the majority of migrant occupations cannot provide a great deal more. Garment workers, for instance, make an average of $1500 per year after over time, of which roughly 49 per cent is remitted to the rural household. Other occupations (with a mean of 36 per cent remitted for all occupations excluding garment work) tend to send back less.

Nevertheless, the vital importance of rural land to translocal livelihoods does not render the modern sector a secondary concern. On the contrary, not only is the regular income obtained from family members' remittances a vital source of funds to purchase key farming inputs during peak agricultural periods, but migrant incomes are also a key means to leverage further funds. As a number of informants reported, there is a greater willingness on the part of microfinance institutions to lend against migrant wages, and against garment workers' wages in particular. Thus, although the absence of collateral is a barrier for some poorer potential borrowers:

> If people have a family member in a garment factory, then they will take out a loan and use the remittances to pay it back every month. It is very easy to get a microfinance loan if you have a family member in a garment factory.
>
> (Focus group, 03/06/13)

As a result of these translocal linkages, the relationship between modern sector income and rural land use is a more direct and explicit one than is generally recognized. Moreover, the impact of these linkages is felt in factories as well as fields. Not only does remitted income provide a key resource for purchasing or replacing agricultural inputs—thereby diminishing the impact of crop loss but potentially requiring higher levels of remittances during peak agricultural periods (Lawreniuk and Parsons, 2018)—but the easy accessibility of microfinance debt to families including migrants constitutes a further pressure on migrant workers from poorer households, who must often meet monthly repayments as well as supporting their own and their rural household's livelihoods. For those unable to match these multiple demands, this may mean the sale of farm land to repay debts incurred through establishing a migrant in the city:

> Before people decide to migrate, [migrants' families] always borrow money from [the preeminent Cambodian bank] ACLEDA for the transportation and to get started. Sometimes this becomes a problem if they cannot find a job to repay. For example, one household had to sell their land because they couldn't find work in the city and needed to repay their debt.   (Phirun, 28/05/13)

Although village level data tell only part of the story in this regard, the evolution of Krang Youv's land market may be observed in Figure 5.4, which uses recall data to

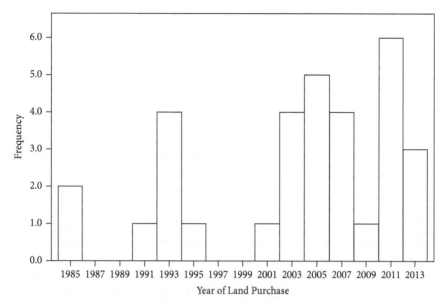

Figure 5.4 Number of land plots sold per year since 1985.

show the number of plots of land sold each year by respondents. It demonstrates both an increasing volume of land sale over time and pronounced peaks in certain years, corresponding closely with some of the most severe nationwide climate pressures experienced during the last quarter century, most prominently in 1985, 1993, 2005, and 2011.

Increases in the volume of land sale have been noted elsewhere as a key impact of climate pressures in Cambodia (Lawreniuk, 2017; Helmers and Jegillos, 2004). However, what has been less remarked upon is how distress land sales such as these feed into a wider system of land use change which rural–urban migration and remitted income play a significant role in catalysing. Indeed, although Krang Youv's larger landowners today have generally been better off than other villagers for decades, the majority of their holdings have been accumulated during the previous one or two decades, by purchasing the land of those forced to sell through debt, ecological shocks or—most commonly—a combination of the two. As smaller landowners reported, 'now the rich have a lot of land' (Farmer, Krang Youv, 16/05/2017), due to bad debts and the necessity of credit:

> Generally, the rich always buy the land from the poor. The poor people are poorer, [so] they always borrow money from the rich and then have to sell their land to them to cancel the debt. Then they rent the land back to them for farming. Now there's a lot of people becoming landless.
>
> (Farmer, Krang Youv, 16/06/2017)

Thus, whilst certain farmers took advantage of acute shocks to purchase land, stating for instance that 'I inherited two hectares, but bought four more between 1986 and 2003 . . . from four or five villagers whose crops had failed for two or three years in a row' (Athit, 28/04/13), longer term shifts away from smallholder agriculture have also proved key. Krang Youv's largest landowner, Phirun (28/05/13), for instance, claims to have accumulated the entirety of his 35 hectares 'from thirty or forty people whose land was too spread out to be convenient for them'. Similarly, a second farmer purchased 3 hectares in 1995 'from two different villagers, who sold their fields because they were too far away from their houses' (Boran, 09/05/13) and yet others stated that the previous owners of their land 'had decided to move away from farming' (Visoth, 23/04/13), or 'to find work outside of the village instead' (Makara, 07/05/13).

Furthermore, the case of Phirun, in particular, demonstrates that with a concerted effort, opportunistic land acquisitions such as these, undertaken over a sustained period, can lead to exponential expansion as the cash reserves generated by previous purchases are used to fund future acquisitions. As landholdings have concentrated, mechanisation has expanded in turn, as farmers well-endowed with land purchase hand tractors or harvesters, or involve themselves in networks 'where we always share each other's machines' (Ponleu , 20/05/13). Consequently, 'the difference [now] is that the rich here have all the equipment for farming, whilst the poor only rent [it] from them' (male beggar, Phnom Penh, 15/06/2017), leading to higher costs and fewer opportunities for those with the least assets. As the pressure placed on smallholders and the landless by these developments have grown, the volume of migration has expanded in turn, to the point that even the reduced supply required in the context of mechanization is no longer available. The consequence is further land concentration and consolidation, as smallholder farmers are unable to meet the cost of labour hire during peak agricultural seasons.

In this way, remitted income is 'funnelled into the pool of household income, which then becomes deployed into the dynamic of production in the village' (Barney, 2012: 75), leading to the ever present inequalities of assets in Cambodian rural life being 'stretched' (Narith, 12/08/13) by the inequalities of opportunity which emerge from them. Translocal livelihoods are key to this process. However, they have not replaced or diminished agricultural livelihoods, but catalysed its redistribution. As villagers explained, it is 'money from rice' that builds the new houses and buys the cars that are the icons of modernity in Krang Youv (focus group, 03/06/13) and it is clear that villagers in all three of the study sites recognize the undergirding role of traditional rural assets and risk in determining the trajectories of change upon which households find themselves. Consequently, 'it seems like fewer people are getting richer. It seems like the families that were rich before are still rich, whilst the poor are still poor' (Bunroeun , Krang Youv, 16/06/2017).

Whilst the idea of a static and homogeneous village production system was always a myth, therefore, trajectories of change are accelerating and moving further apart, as the distinctions between rural and urban disappear. Land and environmental risk, rather than being marginalized in the village economy, remain at its core. However, they have become intertwined in modern sector work, driving the welfare of larger landowners ever upwards through mechanization, land rental and the transition to cash crops, whilst those whose livelihoods were initially less certain, see this historic insecurity hardened into persistent and self-enforcing marginality. Where land and labour once shared a degree of mutual reliance—reflected in proverbs that entreated 'the rich to take care of the poor like the cloth which surrounds you' (Fisher-Nguyen, 1994: 99)—this is no longer the case. As a former farmer, now begging on the streets of Phnom Penh, explained, 'now rich people look down on the poor' (male beggar, Phnom Penh, 15/06/2017), reflecting a dependency that runs only in one direction.

Conclusion

In just a few years since Cambodia's economic reopening in the 1990s, the phenomenon of land acquisition has changed the face of the Kingdom's rural areas. In that quarter century, the country has moved from the rice dominated and largely egalitarian land distribution which emerged in the wake of the *krom samaki* community farming groups during the 1980s, towards a situation in which almost half of rural people are landless or near landless (Sophal, 2008) and cash crops figure with increasing prominence in the rural economy. Thus, although such processes are mirrored elsewhere, ecological change has become 'especially fast-moving in Cambodia, where in just a few years a large area can go from a mixture of forests and smallholder farms to industrial plantation style monocultures' (Davis et al., 2015: 774).

Such rapid changes to Cambodia's ecology and land use have attracted a considerable volume of scholarship in recent years, with both academics and practitioners seeking to interpret the multiple factors that underpin this uneven development. However, 'as important as this research has been, these studies have largely focused on the immediate impacts of the "enclosure" process associated with gaining access to land by investors' (Baird and Fox, 2015: 1). Far less remarked upon has been the translocal and multi-scalar processes by which land use has adapted, in dynamically linked ways, to conditions of rapid mobility, modern sector integration and environmental change, with most research reports and policy work being directed instead towards the forced or induced acquisition of land by large companies (Marks et al., 2015: 33).

By adopting a translocal lens on land use change, this chapter has therefore provided an additional perspective on these processes of ecological change . As it has sought to demonstrate, the translocal linkages engendered by human mobility shape otherwise discrete ecologies in a coordinated manner. Underpinned by a varied and extensive diaspora of mobility, these extend the environmental impact of shifting livelihoods far beyond the area of a single economic concession. Thus, by exploring how patterns of migration dynamically link rural, peri-urban and urban livelihoods it is possible not only to gain insight into changing land use, but also how these processes are intertwined with wider socio-economic systems. Translocal ecologies, therefore, are revealed as fundamentally unequal; who gains and who loses due to the whims of a capricious climate is no longer a matter of chance, but of the ability to mobilize complex and multi-sited economic networks. Transported through these channels of mobility, each drought or flood transforms diverse and distant sectors of the Cambodian landscape yet further, reworking it to resemble ever more clearly the economic inequalities it supports.

6

The Village of the Damned? Narrative, Structure and the Coproduction of Translocal Mobility

The myth of the cursed village of Prey Veng province is famous throughout Cambodia. It is a centuries old story which tells of a community who became highly successful at begging, eschewing manual labour and becoming far richer than their neighbours in the process. In cosmic recompense, the story goes, evil spirits placed a permanent curse on the seekers of immoral alms and all their progeny, so that even today their descendents, rich and poor alike, must leave their homes at least once a year to seek alms, or be struck down by lightning.

Opinions as to the veracity of the story depend upon the tellers, but tend more towards the concrete than might be expected for a supernatural tale of this sort. Some people are able to give specific details as to the whereabouts of the village, recalling a name, a district, or a commune. Others still cite evidence of wealthy people they have seen participating in begging. Moreover, the closer one comes to the various sites of the story, the more specific does it's telling become. Amongst the begging community of Phnom Penh, many see the story as fact, and government officials responsible for the rural areas associated with the myth refute only the supernatural aspects of the tale, admitting that begging by rich people is common practice in many parts of Prey Veng.

As such, rather than describing a single community, the myth of the cursed village alludes to a wider phenomenon of translocal begging migration in Cambodia. Not only Prey Veng, but many parts of Cambodia contribute beggars whose alms seeking in the capital constitutes one element within a broader livelihoods strategy, either on a seasonal or lifecycle basis, when paid employment is impossible or unavailable. Begging migration of this sort, though often characterized by harsh living conditions and persistent rural and urban poverty, should not be viewed separately from the wider migrant system. Rather, it constitutes a sub-system within that system; the informational and practical logistics of which overlap those of labour migrants in places, but are unique in others.

Using qualitative and quantitative data collected in Phnom Penh and Prey Veng, this chapter explores the cultural and normative dimensions of translocal begging behaviour in four parts. Having outlined the conceptual and contextual background to the study, it will secondly examine the logistics and practicalities of

Going Nowhere Fast: Mobile Inequality in the Age of Translocality. Sabina Lawreniuk and Laurie Parsons, Oxford University Press (2020). © Sabina Lawreniuk and Laurie Parsons.
DOI: 10.1093/oso/9780198859505.001.0001

translocal alms seeking, emphasizing the characteristics it shares with labour migration, including remittance behaviour and cyclical patterns. Thirdly, it will analyse the linkages, and overlaps between begging migration and labour migration more generally, demonstrating, in particular, the role of moral narratives in their (co)production. Finally, drawing upon rural fieldwork, it will examine how the structural dynamics of translocal begging serve to reproduce and proliferate the cursed village myth in rural areas, as a widespread and divisive term of denigration.

6.1   Static Perspectives on Begging and Homelessness

Whether in literature or public discourse, analysis of begging is notable for its normative character. Detailed studies are few in number, but those which have emerged, such as Dean (1999), Glasser and Bridgman (1999) and Fleischer (1995) draw upon a rich historical debate on the 'deserving' poor to emphasize, in particular, the moral dimensions of destitution. As these studies highlight, seeking alms is an activity defined and structured by normative discourse. Judgment is not peripheral, but integral to its manifestation, guiding beggars' movements through the carrot and stick of financial incentive and social denigration.

Begging, moreover, is an inherently political issue. As Erskine and Mackintosh (1999) highlight, high level deliberation on the rectitude of seeking alms is centuries old, being attributed to historical personages even by twelfth-century authors. Subsequently, the Elizabethan and Victorian eras were alive with debate over the undeserving poor and how to identify them, even amounting to a compulsory badging scheme designed to indicate those receiving pensions (Hindle, 2004). In recent years, successive British prime ministers, from Thatcher to Blair (Erskine and Mackintosh, 1999), to Cameron have made the theme a favoured refrain within their speeches.

A consequence of this culturally entrenched moral ambiguity is that the worlds of begging and work have tended to remain categorically distinct, even from an economic perspective. Efforts to place Western begging in a wider context have generally tended to do so on a macroeconomic basis, constructing begging behaviour as a response to 'global economic conditions' (Jordan, 1999: 43), but not seeking to link these conditions to the minutia of begging livelihoods in the manner that has so often been undertaken for paid workers. These analyses therefore implicitly assume that the destitute have severed their connection to the global economy.

In the West, where welfare provision has altered the nature of destitution, this model is likely to accord more closely with the realities of homelessness than in the developing world, where studies into the institutional policy (Ramanthan, 2008) of begging and its socio-cultural and religious context (Weiss, 2007) confirm the fine lines which separate begging from work in some cases. Despite instructive early efforts in this context, though, such as Fabrega's (1971) discussion of the relative weightings of stigma and profit amongst Mexican beggars, and Bamisaiye's (1974) exposition of how cultural attitudes combine with the local socio-economy to influence begging practice, studies into begging as a structured behaviour remain relatively rare.

Of those to have emerged in recent years, Krishna's (2010) is the largest in scope, drawing together studies from several countries to highlight how vulnerability places many working people only 'one illness away' from begging. However, other studies in a similar vein adopt an alternative perspective, elucidating, as does Swanson (2010), how begging may act as a 'path to progress' in some cases. Massey et al.'s (2010: 70) study on begging migration in rural India and Bangladesh, for instance, highlights the role of alms seeking as 'a livelihoods strategy' with 'strengths and opportunities' into which people may enter and (crucially) exit. This type of multi-faceted cultural, economic, and social analytical framework is adopted also by Abebe (2009; 2008), with reference to street children, in Adis Ababa and by Kassah (2008: 163) in Ghana, both of which present begging as 'a vibrant informal economic activity', comparable with paid work, but underwritten by powerful cultural norms which shape its economic character.

Nevertheless, as this chapter aims to show, the economic vibrancy of begging is not merely comparable to mainstream work, but intermingled with it. As explored throughout this book, recent trends in mobility have intertwined and overlaid many aspects of contemporary livelihoods, yet alongside other dimensions of non-waged migrant work, begging remains under-explored within the migration literature. Consequently, despite its everyday visibility in Phnom Penh, where slow-paced figures carrying metal bowls are a ubiquitous feature of markets throughout the city, the practice of begging remains enigmatic; the lives and livelihoods of those who undertake it little understood.

Seeking to bridge this gap in understanding, this chapter therefore brings together two periods of fieldwork: June to September 2014; and January to April 2017. It incorporates both rural work, in the villages of Veal, Bo, and Krang in Kampong Trabek district of Prey Veng province, and urban work in the capital, Phnom Penh. Urban data collection took place in eight permanent markets in Phnom Penh: *Psar Kandal* [usually referred to as such], *Psar Chas* [old market], *Psar Thmei* [central market], *Psar Boeung Keng Kang* [BKK Market], *Psar Toul Thom Poung* [Russian market],[1] *Psar Chbar Ampov, Psar Olympic,* and *Psar Orussey.*

Across these rural and urban sites, a total of 43 active beggars participated in semi-structured interviews about begging and migrant livelihoods. These interviews helped to build a consensus that the cursed village was one of a selection of candidates in the Me Sang and Kampong Trabek districts of Prey Veng province.

With the assistance of local informants, the research team subsequently proceeded, over the course of two fieldtrips, to interview a selection of village level and administrative officials, as well as ordinary inhabitants of those villages identified as possessing high levels of begging migration. A total of 10 extended, qualitative, interviews were undertaken across three rural villages, focussing in particular upon attitudes to begging migrants, the curse myth, and the rural livelihoods of begging migrants.

6.2 Translocal Begging Cycles and the Structural Factors that Drive Them

Responses to questions regarding Cambodia's cursed village range in character from vague to emphatic, but most of those asked are able to connect it with Prey Veng province. Moreover, although beggars from all over Cambodia were aware of the story of the village, particulars concerning its location and the nature of the curse itself were strongest amongst those from Prey Veng themselves. As one such informant, a woman of 56 from Ba Phnom district in Prey Veng reported: 'That village is called Veal Ta Prun, in Kampong Trabek district...I have heard from old people that if they don't beg there then they will be struck by lightning, so they have to come [to do it] whether they are rich or poor' (Socheat, Psar Toul Thom Poung, 06/06/2014). A second female beggar, though 'not sure of the commune name' (Kolab, Psar Toul Thom Poung, 08/07/2014) was equally emphatic, stating that:

> I know about that village...That news is real for certain because I have seen many people from Prey Veng asking for money whether they are rich or poor, not like in my village where generally only the poor people beg.
>
> (Kolab, Psar Toul Thom Poung, 08/07/2014)

Although interesting in itself, the myth of the cursed village has a value beyond its own narrative, highlighting a number of important, implicit, attitudes towards begging behaviour in Cambodia, of which two are especially relevant here. First, it emphasizes—behind the veil of the supernatural—an awareness of begging behaviour amongst people who are not poor. Secondly, it highlights a cultural recognition of patterned and circular begging behaviour, rooted in rural areas. Both of these features emerged as pervasive aspects of Cambodian begging behaviour.

Nevertheless, single events, as well as structures, are important. Several of the beggars encountered were victims of idiosyncratic, shocks which prompted a unilinear entry into begging. In particular, Cambodia's bloody recent history has left the country the most heavily mined in the world (ICBL, 2014), producing a huge pool of land mine victims, forced to rely on charity since their injury.

Invalided veterans in particular are a common sight in Phnom Penh, often congregating in tourist areas such as the genocide museum in Toul Sleng and the Cheung Euk killing fields on the outskirts of Phnom Penh in order to solicit alms from wealthy foreign visitors. One such soldier is Rith, a 78-year-old former soldier who explained that:

> I was a soldier, but after the war I got sick and could not work. So I came here in 1984 to beg for income. My family were all killed at the time of Pol Pot... But I have a house in the village. When I am here my nephew stays in it.
>
> (Rith, Psar Thmei, 15/07/2014)

The injuries and illnesses suffered by many former soldiers such as Rith were sufficiently serious as to immediately preclude work in a country which, notwith-standing the prevalence of the issue, provides little or no assistance for those with physical impairments to enter the workforce. As a result, victims of such personal catastrophes have few options beyond begging. Nevertheless, although in some such cases the trajectory into begging has indeed been unilinear—in the sense that the individual in question has exited the labour force and is unlikely to re-enter it—the wider literature on idiosyncratic shocks highlights that this should neither be taken to mean that their connection to the wider macro or micro economy is severed (Swanson, 2010; Krishna, 2010; Dercon, 2004), nor that the unilinearity of their occupational trajectory is reflected in a concomitant geographical stasis (Massey et al., 2010).

Indeed, Rith's subsequent description of the arrangement he has with his nephew hints at a systematic, cyclical process of migration wherein the latter's family takes care of his uncle for the part of the year he returns to the village and is repaid both via the use of his uncle's house when not inhabited and a monthly stipend of '50,000 riel [$12.50] every month to buy rice with' (Rith, Psar Thmei, 15/07/2014). In this way, through a combination of cash payments and payments in kind, Rith maintains a meaningful connection both with his family and rural community, to which he returns on a regular basis.

Nevertheless, the lengthy, thirty-year, duration of Rith's period of begging in Phnom Penh; its instigation via a purely idiosyncratic shock; the uncertain time frames according to which he leaves and returns to Phnom Penh; and even the relative informality of the arrangements he possesses with his relatives, are atypical. Indeed, as the case of Nimol, a 58-year-old male from Kampong Speu highlights, many such begging expeditions are recent or short-term ventures, intended to plug gaps in annual livelihoods:

> I have been coming to the city for about a year... I have a wife and three children, all of whom are married. I still live with my wife, but although we used to have a hectare of rice land, now I have shared almost all of it [with my children] leaving

only 10 Are [0.1 Ha]. I usually do wage labour in the harvesting and planting season, but now it is rainy season, so I have to beg instead.

(Nimol, Psar Olympic, 04/07/2014)

Nimol's example demonstrates how productive labour and alms seeking may co-exist within a household's yearly livelihoods portfolio. However, the biannual cycle he describes fails to fully confer the complexity of his begging behaviour. As he later explained, outside of the harvest and planting seasons which underpin his annual need to beg, his movements are subject to a more rapid cycle of 'fifteen to thirty days spent in my home village and then five to six days in Phnom Penh' (Nimol, Psar Olympic, 04/07/2014). During his time in the city he makes on average 10,000 riel [$2.50] per day, of which he spends around 5,000 [$1.25] on food. After a week, he returns with around $15 to his village. Once it has been spent, he returns again to the city, until the onset of the planting or harvesting season begins to provide work in the village once more.

Far from being unique to Nimol's household, this rapid cyclical begging behaviour was evidenced in a slight majority of cases across the sample (see Table 6.1). Indeed, it appears in most cases to replace the remittance logistics utilised by labour migrants, which more commonly involve money being delivered for a small fee via known taxi drivers (Parsons, Lawreniuk, and Pilgrim, 2014), or—for those with mobile phones—the recently arrived money transfer service, Wing. All of those who regularly migrate between city and country save a (usually fixed) sum of money with them to their rural home, either for their own subsistence or partly to deliver 'to children' or fellow villagers for services such as 'looking after my house when I'm away' (Nimol, Psar Olympic, 04/07/2014).

As shown in Table 6.2, the slight majority of beggars who migrated in a circular manner did so 8.1 times per year on average, alternating alms seeking either with a period of repose in their rural home (more common amongst beggars above retirement age) or rural wage labour amongst those not in possession of rice land. However, most beggars are not landless. Of the study population as a whole,

Table 6.1 Duration and circularity of begging migration

Lifetime duration of begging	Circular migration pattern?		Duration of cycle	Number of cycles per year
(mean number of months)	Yes	No	(mean number of days)	(mean)
52	51%	49%	45	8.1

Source: interview data, 2014. *N*=43.

Table 6.2 Comparative data on labour migrant migration patterns

Occupation	Duration of cycle (mean number of days)	Annual frequency of cycles (mean)
Garment work	86	4.2
Construction work	46	7.7
Motodop	49	7.4
Garbage work	31	11.5

Source: unpublished data from previous study on systems migration in Phnom Penh. See Parsons et al., 2014.

72 per cent of informants possessed a house in a rural area to which they regularly returned, with a further 35 per cent possessing paddy rice land. Amongst those who possessed land, the mean area was 0.25 Ha.

For many of these, a combination of factors—age, economy, and climate in particular—served to render smallholder agriculture non-viable. As they explained, 'I'm getting old, I don't have the energy to do it ... [so] ... I give it to somebody to get the rice back [in a sharecropping arrangement]'. However, though idiosyncratic in part, this decision also formed part of a broader narrative of changing conditions in an agricultural context where 'farming is more difficult now, it's not stable. Sometimes at the beginning of the rainy season there's drought. Sometimes in the middle' (Vichea, Psar Toul Thom Pong, 15/06/2017).

For others, moreover, the annual harvest has collapsed entirely. As a second beggar elaborated, 'Now farming is very difficult. There's no water. This year I didn't get any rice at all' (Malis, Psar Chbar Ampov, 16/06/2017). Yet such agricultural travails were not limited only to those possessing rice land, reflecting instead a landscape of agricultural change which, as detailed in Chapter 4, has shifted to the detriment of wage labourers as well as smallholders (see Figure 6.1.). The oldest beggars, many of whom had never possessed any land but who had nevertheless been able to rely upon a regular, seasonal, demand for their labour in rural areas, outlined this process in lucid detail, noting—as did one 87-year-old pan handler from Prey Veng—the line that divides those who can subsist from their own plots and those, like her, who cannot:

Since I work as a wage labourer, it is difficult now, because nowadays everybody broadcasts, they don't transplant any more. And even when harvesting, they use a harvesting machine, so there's no more job for me ... Now those people migrate outside for work, like they go to a lotus farm. [The people who can't leave] just do their own farming. Previously they used to do their own farming and work for others [also], but now just their own. Generally they all have land, it's only me that doesn't have any land. (Leakhana, Psar Chbar Ampov, 15/06/2017)

Figure 6.1 An expanse of rice fields that stretch as far as the eye can see in Prey Veng province, 2018.
Source: Courtesy of Thomas Cristofoletti/Ruom/Blood Bricks/Royal Holloway.

As such, the 'destitution' embodied by begging behaviour appears to be a transient, rather than a steady state, characterized by a lack of definitional clarity. For example, a number of this study's informants came to beg in Phnom Penh due to a combination of sustained economic and environmental pressures in their home villages, without first selling their land or other assets. Similarly, begging behaviour continues beyond the mitigation of single pressures and shocks, as a durable translocal strategy to plug deficits in annual livelihoods. One such beggar is Socheat, a 56-year-old widow who has been coming to Phnom Penh to beg for over ten years:

> I collect cans and beg in the city, but in the village I also do rice farming. My rice field is 50 Are [0.5Ha] and produces roughly 250 kg of rice every year. This year, though, due to high floods, I lost my entire crop. That's why I'm here at this time. [Nevertheless] I usually come every month, for between fifteen and twenty days. Sometimes I wait until I have saved 300,000 riel [$75] if my health is good, but if not then I have to go home early, often with only 100,000 riel [$25].
>
> (Socheat, Psar Chas, 06/06/2014)

A second case was Chenda, a 38-year-old wife and mother to a family of eight children in Prey Veng, for whom rising waters and falling demand for her off-farm business had made the trip to Phnom Penh an increasingly logical choice. As she explained:

> I used to sell desserts along the road in my home village, but the after three or four years the business failed because of a lack of customers. There were more and more sellers at that time so the customers stopped coming [to me], but I continued to make the desserts every day. Consequently, I got into debt: 100,000 riel [$25] from one man and 230,000 riel [$57.50] from a microfinance institution. I have a large family in my village and my husband works in the fields, but this year the land flooded so there is no work to do ... Our rice land is about 15 Are [0.15 Ha] and makes about 300kg of rice per year but it's not enough for my family.   (Chenda, Psar Kandal, 01/07/2014)

Moreover, this relationship not only between begging and land, but also between begging and business failure, debt, and migration, was echoed by Panha, a 60-year-old woman from Kampong Speu, who related that she:

> previously worked as a palm fruit seller, but got into debt in my home village because the business was going badly. I bought a lot and couldn't sell anything. It lasted around six months but I would lose [money] every day as I needed to buy new stock and nobody would buy anything from me.
>
> (Panha, Psar Kandal, 20/07/2014)

Thus, for some beggars, especially women with families such as Chenda and Panha, begging does not necessarily constitute the end of a sustainable rural livelihood. Rather, in Chenda's case in particular, it is a means of safeguarding the future by avoiding land sale and debt. Panha, by contrast, had 'already sold all' her land but nevertheless undertakes urban begging with a definable purpose: clearing the 'roughly $1000' of formal debt she has accrued as a result of her business's failure by saving some of the '10,000 riel [$2.50] to 20,000 riel [$5.00] per day' she receives from begging.

What the above cases demonstrate, then, is that in addition to possessing a strong cyclical dimension, translocal begging behaviour is in some cases utilized as a means of safeguarding or even improving rural livelihoods. Moreover, by extending analysis to the interaction of begging with urban work reveals an even greater level of dynamism and mutual support, as the case of Bopha, from Kampong Speu, highlights:

> Previously I had no rice land, but we recently bought $250 of land, which is about 50 square metres in size, using remittances of $50 per month from my youngest daughter who works in a garment factory ... She travels to Phnom Penh every day, paying $10 per month to a taxi as it is cheaper than a rented room, although

it takes three hours each way. She leaves at 4am and returns at 7pm. My other four children are married but live with me. Nobody has any land, so they all work as rubbish collectors in Kampong Speu . . . I return there once a week and spend a week [at home]. If there is rain then I will farm. If not, then after a week I will come back [to beg].   (Bopha, Psar Thmei, 25/06/2014)

Above all, Bopha's case elucidates something of the complex translocal diasporas according to which begging takes place. In seeking to derive regular income year round, her household has come to include both daily and weekly cycles between Phnom Penh and Kampong Speu, in addition to the four garbage collectors who travel many miles around the province 'on bicycles and motorbikes' in order to find or purchase sufficient recyclable garbage to sell at a depot at the end of the day. Via this compendium of livelihoods, the family have not only been able to survive both the rural low seasons and Bopha's husband's 'disease', but even save enough to purchase new land of their own.

6.3 The Role of Moral Narratives in the Structure of Translocal Begging

As the above testimonies highlight, begging migration shares many of the translocal characteristics of labour migration. However, the two modes of migration do not simply mirror each other, but meaningfully overlap in complex ways. Amongst the informants encountered during the course of this study, several live with children or other relatives engaged in urban labour of various sorts, as in the case of Phalla (Psar Chbar Ampov, 08/07/2014), who stays 'with my daughter who is a petty trader selling dried bananas . . . [and whose] . . . second child is a construction worker and his wife a garment worker'. Others have children or other relatives working as 'garment workers' (Chenda, Psar Thmei, 01/07/2014), 'as construction workers in Chom Chao [West Phnom Penh]' (Guen, Psar Orussey, 07/07/2014), or even as 'labourers in Thailand' (Chanlina, Psar Toul Thom Poung, 15/07/2014).

As such, unlike the picture of beggars as 'like-situated individuals cut off from wider society for an appreciable period of time' (Glasser and Bridgeman, 1999: 54) presented in the Western context, Phnom Penh's beggars tend to retain strong linkages to their home communities. Indeed, in a manner reminiscent of a number of classic and more recent models of deagrarianization (Rigg, 2012; Bryceson, 1996) and labour migration (Potts, 2010), the majority of informants return on a regular basis to their rural villages. The remainder tend to live with family members working in Phnom Penh, as shown in Table 6.3 and in many cases contribute to household livelihoods alongside workers employed in both the

Table 6.3 Living arrangements of circular migrating and non-migrating migrants

Living arrangements—place of primary residence in Phnom Penh

	Circular migrating beggars (%)	Non-migrating beggars(%)
On the street	50	33.3
In accommodation with family	18.2	38.1
In accommodation without family	31.8	28.6

Source: interview data, 2014. *N*=43.

formal and informal sectors. Only a small minority, however, intended to utilize these connections to enter the labour market.

In most cases, the reason for this was either demographic or physical. Children under eighteen, the old, and the physically disabled [*chen pikar*] would be unlikely to secure modern sector work, even if they were inclined to try to seek it. Indeed, in the latter case, the 'significant exclusion' (Thomas, 2005: 7) suffered by Cambodia's disabled population is particularly strongly felt amongst the labour force. Amongst able bodied people of working age, though, a particular subset of the sample remains: mothers with young children, unable to reconcile the requirements of childcare and work, are a regular feature of Phnom Penh's begging landscape. For instance, Rachany, a woman in her forties explained that:

> I have been coming and going [between my village and Phnom Penh] for ten years. I began working here as a construction worker alongside my husband, but he had an accident and died. He was drunk and got hit by a car when he was crossing the road. My children are small, so I had to give up construction work and do something else. Now I go to the pagoda every two or three days to beg for rice and money...Once my children have grown up a little older, though, then I will go to work as a cleaner in a garment factory.
>
> (Rachany, Psar Olympic, 21/07/14)

Rachany's exit from the labour force and entry to begging is a temporary response to a tragedy which not only slashed the earning capacity of her household, but also generated a practical impediment to her resuming work. To compensate, her two oldest children, aged seventeen and twelve, have taken up work in the village, 'making wigs' and 'washing dishes in the market' (Rachany, Psar Olympic, 21/07/2014) respectively, until they are old enough to shoulder childcare responsibilities for the youngest, a two-year-old boy who travels with his mother during begging expeditions to Phnom Penh. Moreover, in addition to childcare, a second common cause of working age begging is short-term injury, as in the case of Kunthea, a woman of forty:

> I have been begging in Phnom Penh for three months. Previously I worked as a plate cleaner in an Indian restaurant ... for two years after leaving my home village due to high floods ... [Three months ago, though] ... I broke my leg after a car hit me in the street and it still isn't fully healed yet. When it gets better I will look for another job but right now I make 5000–10,000 riel [$1.25–$2.50] per day from begging and sleep on the street, just as I did when I worked in the restaurant.   (Kunthea, Psar Boung Keng Kang, 20/07/2014)

Like Rachany's, Kunthea's case provides a clear demonstration of the fragility of migrant livelihoods. Those who are unable to call upon strong rural–urban networks to assist them in times of need are perpetually only an illness or accident away from needing to beg, sometimes indefinitely. However, as Kunthea's testimony also highlights, the line between labour and alms seeking is not as clear as might be imagined. In common with a significant number of other extremely low-paid workers on the basis of this study, Kunthea's two years of employment at the Indian restaurant rendered her what might be termed working destitute, whereby, as she was unable to afford to rent a room, she slept on the street amongst beggars, 'gangsters taking drugs' and police hassling (Kunthea, Psar Boeung Keng Kang, 20/07/2014).

Even migrants in formal employment, such as the garment industry, are vulnerable to such working destitution if the pressures to their household livelihood are serious enough. As a woman in her forties from Kampong Speu related, health shocks can be especially draining in this regard:

> My husband is in Kampong Speu, but currently has a disease. I had eight children, but two have died. Of the remaining ones five are married, but my youngest daughter commutes to Phnom Penh every day ... to work in a garment factory ... When I'm here, though, I sleep on the street at night. So does my daughter.   (Bopha, Psar Thmei, 25/06/2014)

Combined with the significant incidence of beggars living in rented rooms alongside working migrant relatives, the lack of clear distinction between the worlds of low-paid migrant labour and begging becomes clear. Beggars such as Chenda (Psar Orussey, 07/07/2014) who lives 'near Pochentong [airport] with my aunt who is a garment worker'; Kolab (Psar Chbar Ampov, 11/07/2014) who 'shares a rented room for $80 per month with my granddaughters'; and Phalla (Psar Toul Thom Poung, 19/07/2014) who stays with 'one of my married children, a petty trader' underscore something vital about begging and labour migration: they are not merely similar, but overlapping sectors of the same system.

It is, above all, this indistinct boundary between working and begging which contributes to the culture of suspicion and mistrust surrounding begging in Cambodia. Rumours and opinions that 'they are thieves' and 'will do anything for money' jostle with equally pernicious rumours that they beg out of a desire for 'freedom' and 'to do anything they want' (monk, Wat Koh, 10/08/14). Moreover, as the monk went on to explain:

> We will let beggars stay for two or three days if they ask the monks, but then they have to [move on and] go somewhere else afterwards. Sometimes they go to another pagoda or sometimes they just go home. Mostly they claim to come from Prey Veng, but we cannot know if they are being honest or not. They may just say Prey Veng because there are a lot of poor people there.
>
> (Monk, Wat Koh, 10/08/14)

The monk's statement, perhaps surprising in view of the pastoral role played by the pagoda in Cambodian communities, is indicative of a generally negative, or at the very least sceptical, attitude towards beggars in Phnom Penh and Cambodia more widely. In particular, the conflation of highly heterogeneous income sources, such as 'rubbish collecting, begging and thieving' (monk, Wat Koh, 10/08/2014), indicates how many of Phnom Penh's poorest migrant workers are tarred with the broad brush of vagrancy and delinquency.

These moral narratives are key factors in structuring begging migration. Unlike most other migrants, Cambodian cultural attitudes towards the 'deserving' poor mean that the majority of Phnom Penh's beggars are either below or above working age, widows with young children, or disabled to an extent which precludes access to work. Anyone who could potentially work—even those old, ill or injured to a significant degree in some cases—tends to be treated with suspicion, even by figures of moral authority.

Those who do not fit this profile must generally find alternative income generation strategies, explaining, as did a woman in her forties, that 'I ask the monks at the pagoda for rice every day ... [because] ... I am too young, so people wouldn't give me anything [if I begged on the street]' (Rachany, Psar Boueng Keng Kang, 21/07/2014). Rachany's statement highlights one of the key cultural attitudes underpinning begging in Cambodia: that, in a manner reminiscent of historical standpoints in the West (Hindle, 2004; Erskine and Mackintosh, 1999), to be deserving of charity it is not enough simply to be poor, or even destitute. Rather, a visible impediment to work—generally either age or disability—is a key part of what motivates Phnom Penh residents to give to beggars and therefore a significant determinant on what they earn.

As shown in Table 6.4, the quantitative data reveal a significant relationship (Spearman's, $p = 0.012$) between disability and reported daily income. Moreover, beggars above or below working age (using 18 and 65) earn around 25 per cent

Table 6.4 Earning patterns amongst various migrant categories

Category of beggar	Disabled	Above or below working age	Working age (excluding disabled)	Male (able bodied)	Female (overall)
Median daily earnings	$4.87	$3.75	$3.00	$2.50	$3.75

Source: interview data, 2014. *N*=43.

more per day than working age beggars [$3.75 compared with $3.00]. Likely due to a cultural narrative of female childcare (and the physical presence of babies in some cases), female beggars also earn more than their male counterparts, with able-bodied men of working age earning the least amongst the sample, at $2.50 per day on average, around half of the mean $4.87 earned by disabled beggars overall.

Thus, the charity upon which Phnom Penh's beggars depend is far from indiscriminately meted, but appears, instead, to be allocated according to a narrative hierarchy of desert around which the myth of the cursed village is intertwined. Household wealth is more difficult to measure than age or disability, potentially casting doubt on how 'deserving' potential alms seekers may be, but the suspicion that some of those seeking alms may not be as deserving as they appear is at once confirmed and allayed by the idea of a curse. As such, in Phnom Penh, a supernatural narrative of this sort serves to dislocate the idea of begging from those of saving, circular migration and labour, to which in reality it is tightly tied. This disjuncture between the practical motivations of begging and its moral interpretations are recognised and bemoaned by alms seekers, who complain that 'they still criticise me, but I don't know why. They don't criticise the other migrants. That kind of work nobody criticises' (Malis, Psar Chbar Ampov, 15/06/2017). Similarly:

Some people talk about me, but I don't care, I just come here to pay money and pay off my loan. I don't know what they say [about me], but mostly I don't stay here and when I go home, I never go to other people's houses. I just stay at my house...Now it is very difficult. I have no husband, no children and I'm very sick. I have a lung problem, a stomach problem. All the others are old [too]. Only old people beg, because they don't have any people in their home village to help, so they come here. If people have children, then they send them to work as construction workers, or in Thailand. Nobody criticises them.

(Malis, Psar Chbar Ampov, 16/06/2017)

Notably, moreover, this distrust between beggars and city-dwellers extends also to relationships between beggars themselves, making city life for many migrant alms seekers a solitary affair in which 'nobody in the city helps [anybody]' (Kunthea,

Psar Kandal, 20/07/2014). Indeed, the combination of hostile urban attitudes, low incomes and short-term migratory logistics means that many beggars migrate in pairs or groups from their home village, sometimes with family members and sometimes with non-begging migrants. For example, numerous beggars reported having made their way to Phnom Penh 'with people from my village who work as construction workers' (Chanlina, Psar Chas, 15/07/2014). In addition, as described by Bopha (Psar Thmei, 25/06/2014), part of a trio of beggars from the same village in Kampong Speu:

> the three of us are friends in the village and we all decided together to come to Phnom Penh to work like this. Another friend of ours came first and told me about the opportunities. She worked in a garment factory.

In other cases, the reverse situation is evident, wherein well established groups of beggars from a particular village provide the networks and support for non-begging activity to be undertaken by other villagers. Under such circumstances, the linkages between migrant labour and migrant alms seeking may facilitate access to begging—or aspects of begging such as transport, social networks or residence—from an occupation which has become connected to begging in a particular village. Two members of a nine strong group of regular beggars from Prey Veng outlined just such a story:

> I have been here [in Phnom Penh] for ten years. I was a construction worker for nine years but my energy was getting low, my legs were not so strong and my health was not so good so I began to follow these [beggar women to Phnom Penh] to sell rice wine from my village about a year ago.
>
> (Prak, Psar Boeung Keng Kang, 29/06/2014)

As such, the moral narratives that structure begging behaviour by favouring evidence of old age or disability over rural socio-economic factors encourage an overlap of logistics between waged and non-waged migrants. In circular fashion, this overlap contributes in turn to the suspicion with which beggars are viewed in Phnom Penh, further encouraging them to seek the help of family, friends and other villagers and thereby entrenching scepticism of their deserving status.

In turn, the small-scale narratives of denigration that emerge from such suspicions have a profound impact on the mobility of those who face them. Certain places are sought out for religious or superstitious reasons, with several beggars explaining that 'when I get some money, I will offer some money to the monks' (Leakhena, Psar Chbar Ampov, 15/06/2017) or that 'I often go to the pagoda. I offer some money to the monks to wish that the next life will be better than this' (Malis, Psar Chbar Ampov, 16/06/2017). By contrast, others are avoided, as beggars 'talk to people [in their home village], but only a little', or 'just stay in my house' as they 'don't want to go to the other houses' (Raksmey, Psar Toul Thom Pong, 19/06/2017). Furthermore, mobility may also be affected on a far larger scale, even for those not directly involved in begging:

> Previously, I used to go fishing at the sea in Koh Kong, but my daughter in law was always gambling and my son couldn't make enough money from fishing to pay for that. So with me as a beggar and her as a gambler everybody looked down on my family a lot, so she ran away with another man. Now my son has moved to stay with me. Now I go and come back almost every week – 5 days here and 3 days there – because I have to pay for land rental and firewood [which is] almost $30 per month... Yes, people speak badly about me and hit me, but I just go somewhere else.   (Raksmey, Psar Toul Thom Pong, 19/06/2017)

In and amongst all of this discourse lies the myth of the cursed village, an apparently extraneous component within an attitudinal context more grounded in pragmatism than superstition and tradition. Nevertheless, such narratives are not conceptually discrete from everyday livelihoods. Rather, they are intrinsically bound together in reality, emerging from the cultural backdrop as a means to process new trends and social change. Structure and culture are in this way co-produced by processes of begging migration.

6.4 Climate, Narrative and the Coproduction of Myth

In their concern with beggars' physical states and assets, an element of alms seeking rarely addressed by begging narratives is the role of climate pressures and shocks in engendering begging migration. Prey Veng, alongside four others, Kampong Cham, Kandal, Prey Veng, Svay Rieng, and Takeo, lies in the midst of the Kingdom's most frequently and severely inundated flood plain, making the area one of the most flood prone in the country (Leng, 2014) and historically engendering a high level of vulnerability to their inhabitants' livelihoods. The importance of this context is underlined by the provincial origins of the beggars interviewed for this chapter. As highlighted in Table 6.5, over 81 per cent of the Phnom Penh begging sample originated in one of these five flood-hit provinces.

That rates of begging are relatively high in Prey Veng may seem unsurprising in this light. Yet as the testimonies above serve to highlight, a minority of the beggars interviewed for the purposes of this study were directly driven to alms seeking by floods or other climate events. Rather, in a manner analogous to the broader process of labour migration in Cambodia, the shocks experienced by those who enter into begging migration are subject to 'filtration' (Parsons, Lawreniuk, and Pilgrim, 2014: 1367) by the social environment, which facilitates the logistics of migration.

Table 6.5 Percentages of informants deriving from most flood prone provinces

Province of origin	% of informants
Kampong Speu	23.3
Prey Veng	18.6
Svay Rieng	14
Takeo	14
Kampong Cham	11.6
Other	18.5

Source: interview data, 2014. *N*=43.

Socio-cultural norms and prevailing narratives are therefore key factors in how beggars and potential beggars experience climate pressures and shocks, but the obverse is equally true. Prey Veng's acute and widespread vulnerability to the climate, in recent years especially, has encouraged an upsurge in begging and a concomitant narrative of strong denigration for those who undertake it. Thus, in the Kampong Trabek district of Prey Veng Province, which alongside its neighbouring district Me Sang is the heart of the region associated with the begging curse, the mysticism associated with the begging myth appears to have no place. As the chief of the commune explained:

> There are people here who beg and even if they get a big house they keep begging... [They] are just greedy people who want more and more... So I think that the origin of the superstition is just from people kidding. When people see rich people still begging and they think that they should stop but they don't they may say to them 'you can't stop [begging] or you'll be hit by lightning'.
> (Kampong Trabek Commune Chief, 15/08/2014)

As such, in Kampong Trabek it is pragmatism rather than the paranormal that underpins the narrative of alms seeking: supernatural obligation is replaced by acquisitive greed and the notion of a ritual pilgrimage subsumed beneath more pernicious accusations of highly structured, profitable, migratory begging behaviour. Indeed, references to migratory cycles and systematic household income generation pervade almost any conversation on the subject. Beggars are said to 'go to beg for a fortnight and then come back' (Kampong Trabek Commune Chief, 15/08/2014); 'send back about a million riel [$250] every three months' (Ponlok, Veal Village, 15/08/2014); and to 'go in groups of ten to fifteen people' at a time (Pisey, Bo Village, 16/08/2014). Moreover:

> They treat it is a normal job. They all do it together, sometimes earning about $50 per day. They go everywhere in Cambodia: Mondulkiri, Rattanakiri, Phnom

Penh. Sometimes they meet people who want to come back to see their home, so they take them to a very small, poor, house and then afterwards go back to their nice big house. Working like this they can make about one to two million riel [$250–$500] per week. It's easy work, so they don't want to do anything else.

(Pisey, Bo Village, 16/08/2014)

Although such testimonies, posited by either by non-beggars or those who consider themselves amongst the poorest in their village and hence justified in begging, have the appearance of exaggeration, the highly structured patterns of alms seeking encountered in Phnom Penh and lend these claims a degree of validity. Furthermore, the story was repeated by numerous informants across multiple villages, often in the first person, as one informant, a lady of 77, demonstrated:

I just got back from Phnom Penh yesterday. It is possible to earn 900,000 riel [$225] in a month or a month and a half [there]. I rent a room for $4 per month in Preik Pnou [north Phnom Penh] and share the rented room with my husband who is still working there now. I stay about six weeks and come back just to bring money to give to people I owe it to. I owe $1600 from when I got sick and went to hospital for back pain ... I'm not sure when I'll go back, but my husband will stay until Pchum Ben [the annual festival of the dead, around five weeks after the date of interview]. Normally we stay together in Phnom Penh, though.

(Ieng, Bo Village, 16/08/2014)

As her testimony demonstrates, Ieng feels largely justified in seeking alms due to the large debt she has accrued for medical treatment and the lack of alternative income sources. She and her husband are elderly and impoverished, she argues, and thus have few options beyond their regular begging migrations. In this respect, she is not alone, being joined by various others whose poverty makes then unashamed of their engagement in structured begging. Ponlok, for instance, a landless widow in her forties, explained that:

we call ourselves the begging village because everybody here begs ... I generally beg in the villages around this one, or sometimes this one [itself]. Now I have to stay here because my brother has gone to beg, but when he comes back I'll go again. (Ponlok, Veal Village, 15/08/2014)

However, in this respect, Ponlok and Ieng are unusual. Although begging is extremely widespread and apparently highly systematic in villages such as Bo, where both women reside, it is a source of shame for many. Though something of an open secret, many are reluctant to discuss the practice with outsiders, as noted by many of those who were willing to consent to interviews. Warnings that 'if you go there they won't tell you anything' (Pisey, Bo, 16/08/2014); 'the rich [beggars] won't answer your questions' (Narun, Krang, 16/08/2014); and 'they don't like to talk about it because they don't want people to know' (Ponlok, Bo, 16/08/2014) are offered in response to almost any enquiry. Though not entirely true, these appear to be used as a means to underline the questionable morals of 'undeserving beggars', compounding ill-gotten gains with duplicity and manipulation.

Indeed, it is vital to understand the discourse of begging in Kampong Trabek not only as a passive commentary on a socio-economic phenomenon brought about by Prey Veng's high levels of climactic instability and strong migratory linkages to Phnom Penh, but also as the manifestation of both aggressive and defensive positions within a wider moral conflict. Begging is almost never discussed in the abstract, but is directed towards other parties via accusations, aspersions, and indictments, both within and between villages.

Even Kampong Trabek's commune chief undertook similar behaviour, arguing that it was the Christians within his jurisdiction who are at the root of the problem as 'about fifty percent' of them beg (Kampong Trabek Commune Chief, 15/08/2015). Elsewhere, villages compete in the size of the percentages they attribute to neighbouring communities, asserting, for instance, that 'about ten percent of the people in this village beg, but over there in Krang village about 70% of them beg' (focus group, Bo Village, 16/08/2014). Moreover, as Pisey (Bo Village, 16/08/2014) elaborated:

> Seventy percent is a correct figure [for begging prevalence in Krang village]. About seventy prevent of people beg there. Over in Bo village, there are fewer beggars than here, but when for instance an NGO gives them something to eat, then they just stay home and eat it all, or waste it and then go out to beg again.

In some cases, individuals may even accuse their own village of immoral behaviour in this regard, as did one elderly man living on the fringe of Krang village, who thought poorly of the practice. As he related:

> I'm not a beggar but I do know that a lot of people do begging in this village. Nearly ninety percent of people in this village beg. It is a true story that here even the rich people beg. All of them are just greedy people. They want to get more and more from begging and because of it they can make a lot of money.
>
> (Narun, Krang, 16/08/2014)

As Narun's disapproval highlights, begging is intimately bound up with morality in Kampong Trabek, and with accusations of laziness and greed in particular, but appears to have little to do with any sort of superstition. Indeed, those who were aware of the curse myth generally offered prosaic, rather than supernatural, explanations for its existence, suggesting for instance that 'it's just a way for people to look down on beggars' (Kampong Trabek Commune Chief, 15/08/2014). However, there is also a suggestion that the myth has been appropriated by its 'undeserving' targets, as several residents of Bo village agreed:

> That story has never [actually] happened here. It is just a thing that people say to each other ... because they want to look down on beggars who have made some money but keep begging. They beggars who are rich also say it [about themselves] though, because otherwise people round here wouldn't give them any money.   (Focus group, Phnom Penh, 31/07/14)

Such testimonies show that, rather than being merely a part of the cultural backdrop of Kampong Trabek, the cursed village myth is very much alive and in use as a narrative of active animosity both within and between villages in the area (see Figure 6.2.). Individuals accuse other individuals of 'undeserving' begging behaviour, households other households, and communities other communities, always seeking to shift blame for what is viewed as a shameful practice elsewhere. Only the poorest eschew the moral and narrative dimensions of begging behaviour in favour of emphasizing its pragmatic necessity and the need, simply, 'to get something to eat from day to day' (Ieng, Bo Village, 16/08/2014).

Moreover, by incorporating the practice into the wider context of translocal migration, a final, pragmatic light may be shed upon the continued prevalence of Cambodia's cursed village myth. Far more than just a story, it is an active, culturally mediated, commentary on a contemporary situation; a narrative revived from folklore to explain a migration system rooted in ecological vulnerability and economic deprivation. As such, the moral narratives surrounding begging exist in active discourse with its structural dimensions, shaping what has become a major off-farm source of income in the region. These stories therefore direct mobility, but also they also determine its fruits, ensuring that income derived from begging comes at the cost of social sanctions. Through denigration, rumour and exclusion, seekers of alms are distinguished from those who 'earn' income by other means. As in the world of waged work, therefore, inequality amongst even the poorest and most marginal migrants is sustained and entrenched through narrative.

Conclusion

The myth of the cursed village of Prey Veng is instructive to scholars of begging behaviour, and migration more generally, not for its supernatural intrigue, but for what it reveals about the co-production of structure and narrative in translocal communities. As demonstrated through informant testimonies, cultural narratives—even ancient ones—are not discrete from contemporary mobility, but may be invoked and re-popularised to explain new phenomena. Thus, rather

Figure 6.2 A village in Prey Veng province, with houses in the foreground and rice fields extending to the horizon, 2018.
Source: Courtesy of Thomas Cristofoletti/Ruom/Blood Bricks/Royal Holloway.

than being merely a term of denigration, the cursed village myth is an active discourse, invoked by givers and seekers of alms alike, as a means of rationalizing an increasingly prevalent practice. Via this discursive mechanism, Cambodian notions of the 'deserving' poor are both re-ordering and being re-ordered by changing economic imperatives.

Begging migration therefore follows largely similar patterns to other translocal forms of forms of migration: it is neither a descent into destitution, nor a meandering and aimless strategy. Rather, it is a heavily stigmatized, uncomfortable, and ill paid form of employment whose main advantage is accessibility to the less able bodied. Like other translocal earning strategies, it is characterized by circular and chain migration patterns; involves remittances which may either be invested, consumed, or directed towards the service of debt; and it is managed in many cases according to a multi-occupational household strategy. Consequently, begging migration in Cambodia does not only arise in an ad hoc manner, or on an isolated household basis, but—like labour migration—forms part of a persistent system of networks and linkages which influences the choices of those involved.

Key to understanding the production of translocal discourse, though, is that this systematic element and the re-emergence of the cursed village narrative are inextricably intertwined. Translocal patterns of labour—instigated by socio-economic and environmental stressors—have not only sparked the proliferation of the story, but shaped it. The myth of the cursed village is interpreted quite differently in rural and urban areas: practically and underpinned by denigration in one; and more supernaturally and laced with moral judgement in the other. Yet both forms of the narrative combine to influence the mobility of its subjects, who in turn structure its reproduction in both areas. This revitalised folk tale is therefore neither rural, nor urban, but exists amongst and between these spaces; a story born of and sustained by mobility.

Notes

It should be noted that these are not a direct translation of the market names as they are known in Khmer. For instance, the large, art deco, 'Central Market' is known to Khmer speakers as *Psar Thmei*, or New Market. Similarly, *Psar Toul Thom Poung* is referred to by English speakers as The Russian Market, whilst the direct translation would be 'Market on a big hill'.

7

We Move Therefore We Are

Cambodia's Translocal Politics of Nationalism

Always a fiercely proud country, commentators have in recent years noted the re-emergence of a Cambodian nationalism rooted in ethnicity and exclusionary politics (see e.g. Frewer, 2016; Vannarith, 2015). Fuelled by the populist narratives of the opposition Cambodian National Rescue Party [CNRP], which briefly threatened Hun Sen's power prior to its effective dissolution in 2017, 'anti-Vietnamese political rhetoric...has gained steam since the general election in July 2013' (Vannarith, 2015: 1). In the intervening years, this rising tide of ethnic prejudice has transcended the boundaries both of politics and formality, as opposition speeches 'frequently employ slurs to refer to the Vietnamese and pin a lack of jobs and some crime on their presence in the Kingdom' (Taing, 2016: 1), government authorities deport record numbers (Ros, 2017: 1) and violent attacks on ethnic Vietnamese grow in number and severity (Millar, 2016: 1).

Elsewhere in the world, the revival of populist politics during the twenty-first century has been characterized as a rejection of globalization (Peters, 2018; Fotopolous, 2016), but in Cambodia the opposite appears to be true. During the past two decades, global and regional integration have driven 'remarkable economic growth' in the Kingdom (Hokmeng and Moolio, 2015: 6) and recent national minimum wage negotiations have seen garment workers appeal to, rather than reject, the international order. At the same time, nationwide support for the $177 minimum wage campaign—which bridged both geographical and occupational divides with great effectiveness (Lawreniuk and Parsons, 2018)—served to demonstrate the translocal nature not only of livelihoods but solidarity.

As the ruling CPP seeks to prevent a repeat of the opposition CNRP's success in the 2013 and 2017 elections, where strong electoral gains forced a nakedly authoritarian response from the government—dissolving the CNRP as a party, imprisoning or exiling opposition leaders and forcing the closure of a number of media outlets and NGOs—mobile livelihoods have taken centre stage in Cambodian politics. Migrants' relationships to each other and the wider world have become integral to the production of nationalist narratives, as all sides have endeavoured to affirm the unity of the nation in relation to an external—and internal—other. Nevertheless, despite allusions to their interconnection, the twin processes of nationalism and translocality have yet to be examined in terms of

Going Nowhere Fast: Mobile Inequality in the Age of Translocality. Sabina Lawreniuk and Laurie Parsons,
Oxford University Press (2020). © Sabina Lawreniuk and Laurie Parsons.
DOI: 10.1093/oso/9780198859505.001.0001

their linkages and mutual influence. This chapter will do so, highlighting first how recently shifting patterns of economy and mobility have altered the nature of national politics, before demonstrating the influence of this new domestic politics on the key victims of political exclusion: the Vietnamese ethnic minority that make up 5 per cent of the population.

7.1 Conceptualizing the Nexus of Mobility and Nationalism

After years of losing the battle for the hearts and minds of Cambodia's rapidly swelling migrant workforce, Hun Sen has taken the flight to the factory floor. Not content with a long-standing personification of government subsidies allotted to the garment industry in each wage negotiation—colloquially referred to as 'Hun Sen money' by workers—the prime minister has in recent months allotted each Wednesday afternoon to meeting some of the 1,000,000 employees (ILO, 2018) of the nation's premier exporter. Mirroring the CNRP's successful social media strategies, Cambodia's three decade incumbent ruler is now to be seen on Facebook ironing fabrics, greeting crowds of workers and distributing envelopes of money around the capital. Migration, on this evidence, has moved to the heart of the political agenda in the Kingdom (Lawreniuk, 2019).

Nevertheless, the political importance of mobility is no inspiration of Cambodian politics. Movement has long been recognized as breaking down geographic and interpersonal identities even as it re-draws others. Around the world, the 'heavily debated' issues of assimilation and integration (Schneider and Crul, 2010: 1143) are a key example of this, drawing mobile conceptions of belonging, nationhood, and ethnicity to the heart of global political discourse. From the Mexican border wall to Brexit, the world is acutely aware of the political power of movement, yet there is bias to this awareness. The evolution of '"new" forms of identities... "pluralistic" or "multicultural" societies, and erosions of myths of homogeneity' (Tesfahauney, 1998: 501) have presented ample examples of how migration and identity are linked, yet these are invariably underemphasized, masked by the widely received assumption of movement as a source of disjuncture and dislocation.

The issue of 'methodological nationalism' has played a key role in this respect (Kalir, 2013: 311). The majority of studies retain the crossing of national borders as a precursor to investigation, thereby excluding domestic mobility from analysis. Not only does this 'privileging of state borders and categories in many of the mobilities studies' (Kalir, 2013: 311) leave much of relevance unexamined, but it also invokes a damagingly circular reasoning. 'Narratives of movement can actually endorse the normality or historicity of stasis' (Glick-Schiller and Salazar, 2013: 3; see also Pershai, 2008; Lawrance, 2007), but if nations are seen as the categorical basis of investigation, then their validity in that role tends to proceed unquestioned. How mobility feeds myths of permanence and stasis therefore remains a question rarely asked and far less answered.

Nevertheless, it is a crucial one. Counter to the mythmaking discourse of most national narratives, the nation state itself is a product of mobility, evolving in particular out of the rootlessness that characterized the industrial revolution (Ringmar, 2016), during which some 85 per cent of the European population relocated their home (Schultze, 1998). At a time when the agriculturally mediated ties between labour and land were being severed to an unprecedented degree, 'the nation was the new home of the homeless and the new earth in which they were required to reroot themselves' (Ringmar, 2016: 7). Thus, it was to these 'existential worries that the idea of the nation provided a response' (Ringmar, 2016: 7) and this connection has retained a high relevance in the intervening years. As Tesfahauney (1998: 501) outlines:

> On the one hand, the international mobility rights of individuals are structured by and integrated to the sovereign prerogatives of nation-states to control their boundaries . . . At the same time, international migration is an expression of or contributes to the erosion of various conventional and nation-state based economic and social boundaries.

However, whilst the shoring up of statehood and sovereignty through mobility (e.g. De Genova, 2002: 425) is a topic of interest in itself, its linkages to inequality are yet more pertinent. The stratified discourse of migration—most famously exemplified by the rhetorical distinctions separating migrants, refugees, and expatriates (Volpicelli, 2015)—is closely linked not only to global power structures, but national ones also. Who wins and who loses as a result of migration is, at least in part, a question of narrative and categorical distinction, the production of which occurs at multiple scales and is linked into broader processes of national identity. Similarly, inequality, national pride, and migration exist in a mutually dynamic relationship (Han, 2013), so that the boundaries of inclusivity within any one depends on the structures mapped out in the others.

Thus, just as 'mobility always entails a relational dimension', unable to exist without reference to immobility (Gerhartz, 2016: 86), so too is nationalism built inherently on exclusion. Yet whilst a boundless literature on the 'dark side of the nation' (Banerjee, 2000: 1) has been complemented in recent years by a growing attention towards the 'darker side of hypermobility' (Cohen and Gösling, 2015), these twin features of identity have rarely been considered in concert. This chapter will explore these inter-linkages, demonstrating how multi-scalar processes of mobility have contributed to nationalistic and exclusionary discourses both within and outside Cambodia.

7.2 The Myth of Stasis and the Birth of a Mobile Nationalism: Crafting a Movement from Movement

The '"changelessness" of Cambodian society' (Chandler, 2009: 10) has been so frequently noted by Western commentators, as to have become something of a ubiquity in any piece of writing directed at describing the nature of a Kingdom in which, it has long been assumed, 'traditions persist, continuity is maintained, and the deep beliefs of the people abide' (Macdonald, 1958: 134). Indeed, this tendency is so ingrained that where any concession is made to the nation's mutability in the face of a thousand years of global dynamism, it tends to be one suggestive of a slow, almost imperceptible decline—a double-edged recognition that whilst it may be 'a smaller, less important place [...] In other ways it is the same' (Macdonald, 1958: 134).

Nevertheless, this view, propounded most prominently by French colonizers eager to present themselves as the instigators of progress (Chandler, 2009), bears little relation to Cambodia's turbulent and mobile history. Rather than stasis, the story of Cambodia has for centuries been one of movement: 'the capture, displacement and resettlement of populations; the raising of armies; flight from invading troops; and the abandoning of cities and villages' (Edwards, 2007: 426). That Cambodia in the early 1990s had acquired a reputation for stasis speaks to the extent of the control exercised by the People's Republic of Kampuchea [PRK] during the preceding decade. Having reached an unhappy apex of forced migration and flight during the 1970s, a nation always charac-terized by 'great mobility' (Kalab, 1968: 525) now had immobility thrust upon it, as an uneven and unregulated economy (Gottesman, 2003) and barriers to movement, during which 'travelling from one province to another, or even within a province was also not without restriction' (Deth, 2009: 76) took hold. The 'tyranny of proximity' (Edwards, 2007: 424) instigated by the French colonists who sought to undermine traditional power bases by technologically extending the reach and power of the state was therefore placed, for a decade, in reverse.

Indeed, so burdensome was the enforced immobility of both Democratic Kampuchea and the decade long PRK period that followed it, that by the time Cambodia's first democratic elections signalled the beginning of an era of unpre-cedented economic openness, the issue of motility was permeated with national-istic significance. Unionization in Cambodia's nascent garment industry not only occurred rapidly and extensively, but in a manner intertwined with sovereignty. Rather than issues of pay, the frequent strikes of the industry's first decade related to poor treatment by managers and supervisors in the form of forced overtime, verbal abuse and 'racist slurs' (Hughes, 2007: 842) and were couched in 'the language of national pride and the rights of Khmers to be treated respectfully in their own land' (Hughes, 2007: 842).

Nevertheless, though underpinned by nationalism, the issues faced by Cambodian unions in their early years were essentially local ones, involving little coordination between the industry's early federations. Only later, once the launch of the ILO's Better Factories Cambodia programme in 2001 and subsequent signing of the US–Cambodia trade agreement in 2006, created significant incentives for the industry to adhere to national and international labour standards (Kolben, 2004), did the landscape of unionism began to shift in character. Both 'a rapid proliferation of trade unions in Cambodia's garment industry' (Arnold, 2013: 13) and an even faster upturn in the number of strike days lost to industrial action (Nuon and Serrano, 2010) combined to engender a sense of shared narrative across the industry. As the sector grew and real wages inexorably declined, therefore, the tinder of a new mobile nationalism simply awaited the spark of leadership.

This would arrive on 19 July 2013, a day which marked the beginning of a shift in the course of the Cambodian political narrative. Returning from three years of exile in Paris, following a 2010 conviction on vandalism charges that he argues were politically motivated, long-term opposition figure Sam Rainsy was greeted by tens of thousands of supporters lining the road the road to the airport (see Figure 7.1). The political enthusiasm generated by his presence, which followed a

Figure 7.1 Opposition party co-leader, Sam Rainsy, addresses a crowd at a post-election rally on Cambodia's riverfront, 2013.
Source: Courtesy of Owen Taylor.

late pardon granted by Prime Minister Hun Sen to allow Rainsy to participate in the forthcoming national elections, helped to unite and embolden a growing opposition movement led by the newly formed Cambodia National Rescue Party [CNRP]. From Rainsy's return until the election took place on 28 July, the streets of Phnom Penh resounded to calls of *Doh min doh?* [change or no change?] from passing cars, as bystanders, pedestrians and motorists called back *Doh!* [change!] in ever more confident tones.

Ultimately, the election itself—marred by widespread electoral irregularities (ERA, 2013)—returned a reduced majority for Hun Sen, whose Cambodia People's Party [CPP] won 68 seats against 55 for the CNRP. Even before the party's forcible dissolution, tangible legislative gains from the presence of a united opposition were somewhat limited (Future Forum, 2016) and as Caroline Hughes (2015) points out, the public prominence of the CNRP belied a level of opposition support that has remained roughly consistent during the last two decades. Nevertheless, if the formal political impact is equivocal, the social impact has been remarkable, as the CNRP's harnessing of Cambodia's growing union movement has politicized industrial solidarity in a manner that has reshaped civil society in Cambodia.

Indeed, the mass strikes of December 2013 and January 2014 that followed the disputed election result were neither purely economic, nor merely a bubbling over of anger at an election many CNRP supporters viewed as 'stolen' (Brinkley, 2013: 1). Rather, as Arnold (2017: 31) argues, these protests signalled the rise of a new kind of politics in Cambodia, wherein 'workers' autonomous actions mark the limits of their participation in trade unions/civil society terrains, signalling an emerging political space beyond the integral state'. Otherwise put, both the CNRP and the union movement—or, at least, the minority of which that is either explicitly opposition supporting or independent (Arnold, 2013)—have recognised the vast potential power of channels alternative to mainstream politics, to achieve both parties' ends.

What distinguishes this new centre of political gravity above all, is its predication on mobility and mobile livelihoods. More than 63 per cent of the Cambodian garment industry are inter-provincial migrants (CARE, 2017), but to consider them in this sense alone is to misrepresent the situation. They are not defined by *having moved* but by their *ongoing mobility*, meaning that 'ever larger numbers of the population do not fit the mould for which CPP strategy was designed' (Hughes, 2015: 13). Rather, theirs is a translocal politics, engendered and structured by the demands of a livelihood that incorporates both rural and urban elements in dynamic interplay. As workers explained, what drives them to protest their wages is not urban hardship, but poverty manifested across multiple locations:

> If the family in the home village is very poor, then it is very difficult for them to
> send enough because they need so much. My family is average, so it is OK for me,

but it is very difficult when there is no rain because I need to send a lot of money...Workers from poor families have a very difficult life because they send everything home and then have nothing left.

(Female garment worker, 22, 15/07/2015)

Given these close linkages between agriculture and urban work, the role of agricultural pressures emerges as a key factor in fomenting urban protest. Indeed, this is reflected in the testimonies of a number of workers, who cite the pressures of sustaining their family farm as paramount in encouraging them to join the protests of 2013 and 2014. Thus, in a country historically riven by deep rural-urban divisions (Davies, 2010; Chandler, 2009), a new form of grassroots solidarity has emerged out of the growing number of people who do not conform to traditional socio-political boundaries. An identity forged at multiple scales, this constitutes the genus of a nationalism dominated by a global economic outlook, but fuelled by regional prejudice and historical fears.

7.3 We the Workers: National Solidarity in the International Market

The wave of opposition stoked by the CNRP in the run-up to the 2013 general election did not break on the day of the result. Rather, fuelled by perceived injustices and irregularities in the voting system, opposition leaders refused to take their seats in parliament, organized large-scale protests and courted media coverage of their discontent. At the same time, the newly galvanized and coordinated union movement took to the streets themselves in protest against the government's announcement of a mere $15 rise in the minimum wage—from $80 to $95—demanding the $160 wage mandated as necessary by a recent report on the industry (Kashyap, 2015). Moreover, these two axes of discontent were intimately and explicitly intertwined. Leaders on both sides pledged mutual support and attended one another's rallies. Striking garment workers attended CNRP protests (Lipes, 2013) and the frequency and intensity of industrial action grew, until on 4 January 2014, police opening fire on protestors, killing five protestors and instantly scuttling the mood of national protest (see Figure 7.2.). As a garment worker living nearby reported,

I heard the gunshots and saw them carrying the wounded along here, to help them in the houses. There are some people missing even now, many people died...The government sent the soldiers in to attack the protestors. They hit them, which made the protestors angry. So they began to throw stones and burn things like tyres.   (Female garment worker, 09/07/2015)

Figure 7.2 Police and demonstrators standoff along Phnom Penh's riverside as tensions rise following post-election protests, 2013.
Source: Courtesy of Owen Taylor.

The impact of the government's bloody crackdown was immediate. Although wage negotiations were subsequently resolved for a (still) record rise of $28—an additional 28 per cent rise on the $100 minimum agreed in 2013—many unionists viewed their movement as having 'failed' in their standoff with the government and garment industry (Operational Manager, Coalition of Cambodian Apparel Workers Democratic Union, 02/10/2015). The lethal crackdown of 4 January had increased the stakes of protest beyond what many protestors deemed within their capacity and although strikes continued, they were diminished in number.

Indeed, as shown in Table 7.1, the number of strikes fell sharply in 2014 compared to the previous year and declined once again the following year. Meanwhile, the number of days lost to strikes fell at an even faster rate, indicating declining strike attendance by workers. Worker testimonies suggest that much of this decline was attributed by workers and union representatives to the atmosphere of fear—both for personal safety and family livelihoods—that entered the union movement following the crackdown. As a garment worker involved in industrial action at the time related, for example:

Table 7.1 Strike Data from 2011 to 2015

Year	Number of strikes	Number of days lost to strikes	Days lost per strike
2011	34	139,513	4103
2012	121	542,827	4486
2013	147	888,527	6044
2014	108	526,944	4879
2015	118	452,364	3833

Source: GMAC.

At that time, people were very afraid. As soon as the demonstration became quieter, they escaped and went back to their homeland. Those who joined were overwhelmed [*kroboul mok*]. They had no choice.

(Female garment worker, 09/07/2015)

As such, protestors past and present agreed that participation had shrunk following the government's crackdown, with recent actions attracting 'only 50% compared to the number before the strike' (garment worker, 21/02/2015). However, the more fundamental difference was in coordination. Having reached a high water mark of national collaboration in late 2013, the aims of strikes and protests were once again mediated at the local level and demonstrations—which had previously been used to generate pressure in relation to issues at the national scale—diminished sharply in number. As a garment worker explained, 'everything is now getting less strong. They still have protests, but not like before. Now they just demand money for food and some small things' (garment worker, 15/07/2015). Similarly, as a union official went on to bemoan, 'since that strike, the factories don't join together, [they just protest] case by case. If each factory has a problem, then they protest about that' (Branch level union representative, 13/07/2015). Both workers and representatives therefore continue in many cases to view the 2013/14 demonstrations as the high water mark of protest in the country, as an official of the Free Trade Union explained:

[The protest of 2013 and 2014 was] the biggest demonstration that we have ever done in Cambodia. Since then, there have been no really big demonstrations. We have called strikes at individual factories, but no demonstrations.

(Free Trade Union Official, 15/07/2015)

Nevertheless, although the numbers attending strikes and demonstrations continued to decline into 2015, the return of union representatives to the negotiating table in 2015 revealed that this fragmentation was temporary. During talks aimed at securing a $177 minimum wage, it became apparent that rather than under-mining the national solidarity awoken by political circumstances in 2013, the violence that greeted the start of 2014 had led to a hardening of mindset in certain quarters. Those who continued to protest explained that their certainty of entitle-ment to better conditions allowed them to overcome their fear, stating, for example, that 'I am now no longer afraid to protest because I am only asking for what I deserve' (garment worker, 13/07/2015). Moreover, a common theme in the protests in 2015 was the debt of gratitude owed to those who had risked themselves previously and duty to continue their struggle through protest. As one garment worker articulated:

> I have been back to protest two or three times since the violent one. I'm not scared because I saw people sacrifice [themselves] in order to get a high salary for me, so I have to do it to compensate those who did it for me.
>
> (garment worker, 15/07/2015)

This sense of national solidarity pervaded both the $177 movement itself and the retrospective narrative surrounding the strikes of 2013 and 2014. As protestors explained, 'at the time of the strike, we were protesting for the whole country' (garment worker, 22/02/2015) in the sense that 'all of the unions joined together to do something for the general benefit of the country', rather than workers within the industry alone (garment worker, 24/07/2015). However, unlike the 2013 demonstrations, the $177 movement saw this narrative reciprocated in real time, as a raft of unions representing workers from other industries became directly involved in campaigning for improved wages in the garment sector. Members of groups as diverse as the farming communities of Kandal, roadside petty traders, and KTV [karaoke bar] workers expressed strong opinions that the garment workers' struggle for fair wages was closely connected to their own. Only through mutual support, they argued, would a better Cambodia be achieved:

> I heard about the $177 movement from the union, so I joined a demonstration outside the Ministry of Labour. Every sector was there: KTV girls, beer girls, everyone. We want to help the garment workers like this so that one day they may help us in return. We are a very small group [otherwise], so we don't have any power.   (KTV worker focus group, 06/10/15)

Even more fundamentally, the age old rift between the politics of the urban and rural, characterized by 'a long history of rural–urban imbalance and, often, antagonism' (Lawreniuk, 2017: 204) has been partially bridged by the translocal nature of earnings for a growing proportion of households in Cambodia. As a farmer from Kandal explained:

It's different from the past. In the past, only one member [of the household] would work and everybody would be provided for. Nowadays, everybody works, but it's still not enough. Remittances [from the factories] are important, but it's not enough for people here to live... [Consequently] ... I encourage all of my children to participate in this campaign. I don't worry about the violence, because this campaign is for everybody, the whole country.

(Coalition of Cambodian Farmer Community
community organiser, 01/10/2015)

This unified attitude towards the greater national good, tinged with a pragmatic acceptance that the size and importance of the garment industry makes it by far the most powerful actor in any potential negotiations, has underpinned the involvement of a wide range of industries in support of the garment workers' protest. However, it also reveals a broader shift in emphasis away from local and industry specific issues towards protest as a means of 'becoming active members of the body politic' (Arnold, 2017: 23). Emboldened by the legitimacy of a document which supported their claims, protestors have begun to think beyond the narrow confines of union negotiations and to place their case in the context of the global economy, appealing directly to buyers to demand improved wages on their behalf.

In contrast to the $160 campaign of 2013, which largely focused on the targeted minimum wage figures themselves, the $177 campaign adopted an international perspective. Replacing the Khmer language placards stating simply '$160', or 'we need $160', which characterized the previous campaign, demonstrators two years later brandished signs and shouted slogans aimed directly at international buyers. Some of these were more general in scope, such as the message, in English, that 'THE BUYER MUST PROVIDE BASIC WAGE $177' alongside brand logos including H&M, Adidas and Gap (Teehan et al., 2014), whereas others targeted specific companies, with signs asserting in places that 'Zara starves Cambodian workers' or 'C&A starves Cambodian workers'.

The early results of this strategy appeared positive. Nevertheless, although initial talks yielded tentative support from several brands, the requested pressure was not placed on manufacturers, leaving industry leaders in a position of relative strength in negotiations. In particular, their greater insight into the economics of the industry allowed discussions to be moved into an arena in which few union leaders were comfortable. As one such leader complained, they felt able to speak 'only on daily expenses, whilst the manufacturers talk about the economy, productivity, etc.', leading the manufacturers to seem 'very powerful this year', whilst the unions 'seem very young' in comparison (Pav Sina, National President, Collective Union of Movement of Workers, 02/10/2015). Above all, however, it highlights the failure of global buyers to make good on their purported support for the campaign. Indeed, as Pav Sina continued:

> Truly, the demand for $177 is not really from the buyer. They are just like the manufacturers, just talking. They say they will buy [garments at a higher price], but then they don't place the orders. We have already made a general letter and submitted it to the big buyers. They replied and said [they support us], but it's just talking. When they order from the factories, they don't increase the price. Sometimes they pay even less than before. Then when we ask the factories [to increase wages] they reply, 'How can we give more salary to the workers if the price of the product remains the same?'

Nevertheless, despite difficulties encountered in the negotiations, which ultimately led to a wage increase of only $12 to $140 during 2015 (Arnold, 2017), the seeds of a more international outlook had been sown. Rather than protesting against the greed of domestic elites, demonstrators have come increasingly to recognize that their well-being depends less upon the factory bosses that they themselves can see (however rarely) and to a far greater extent upon multi-national companies and Cambodia's place within the global economy. More than the threat of being fired for their activism, this realisation has intertwined recent attitudes to protest with an awareness of the potential consequences of buyers' departure from the Cambodian industry. Indeed, as a representative of the union explained:

> Before the demonstration got a positive result: salaries were increased, but we also got a negative one: some investors left the country. So some workers are thinking 'if we protest again, we will lose our jobs'.
>
> (Branch level union representative, 13/07/2015)

In the context of demonstration, this sense of the communal risk attached to protest—both physical and economic—has entrenched the mindset of national solidarity that accompanied the 2015 demonstrations, so that even workers who claim that they 'joined for my own benefit' argue in turn that 'it would not just be good for me, but for my whole family. I work to support the family, so if my salary is increased, then maybe I can send back a lot of money' (garment worker, 21/07/2015). Thus, by 2015, the union movement had transitioned from one dominated by issues specific to the garment industry, to one undergirded by a broader, translocal agenda. Inter-linkages between livelihoods are an increasingly important part of protest and the $177 movement garnered support across the country, as unions representing farmers, teachers, KTV workers and *tuk-tuk* drivers amongst others all joined in active support of the garment workers' protests.

The resulting sense of national solidarity has helped to restore some strength to unionism in Cambodia. As demonstrators proclaimed, 'first, I was very afraid after the violence, but now I am not. If there was a demonstration, I would join it... At that time, the workers were made weak, but not anymore' (garment worker, 16/07/2015). Nevertheless, as with any nationalistic project, inclusivity is meaningless without it corollary and the newly politicized union movement is no exception. By focusing on the livelihoods of Vietnamese and ethnically Vietnamese Cambodians living in Cambodia—and the narratives that surround those livelihoods—what follows shall highlight the inter-linkages of this industrially driven nationalism with rising levels of prejudice and exclusion in national discourse.

7.4 You and the Yuon: Racism and Hate Discourse in National Politics

Amid the early enthusiasm which greeted Sam Rainsy's return and with it the reinstatement of something approaching democratic pluralism in Cambodia, certain elements of the opposition's political message received limited attention in the international press, subsumed beneath the historic scale and excitement surrounding the event itself. The *Phnom Penh Post* described a 'hero's welcome' (Seiff, 2013: 1) and 'local media heralded the homecoming as possibly the largest opposition rally the country has ever seen' (Campbell, 2013: 1). The CNRP leader's statement that 'I have returned to rescue the country' (Campbell, 2013: 1), similarly, received widespread coverage, whilst the fuller details of his discourse—spoken invariably in Khmer—remained hidden. Indeed, as Rainsy elaborated in 2013:

All compatriots – this is the last opportunity, if we don't rescue our nation, four or five years more is too late, Cambodia will be full of Vietnamese, we will become slaves of Vietnam.   (Rainsy cited in Hutt, 2016: 1)

In the wake of the general election, however, a political movement that had presented itself as fundamentally inclusive, calling for national unity in order to move beyond 'the system of inclusion and exclusion that the CPP has, largely single-handedly, imposed on Cambodian politics' (Hughes, 2015: 6) was revealed upon closer inspection as equally, if differently, divisive. Throughout the campaign and its aftermath, references to exploitative, larcenous, and invasive actions of 'the Yuon'—a much debated Khmer term for the Vietnamese, generally agreed to be pejorative in nature—and their supporters grew in number and volume, as exemplified by the following statement, made upon Rainsy's return in 2013:

'We will usher in a new era in Cambodian history to write a new page on the protection of its territorial integrity,' he proclaimed, continuing 'Many Yuons have come. They move their border posts close into our territory.'

'I pity Khmers very much. They have lost their farmland, because Yuons are always coming in, and the authorities do not protect their fellow Khmers at all, but protect the invading Yuons. Now they have brought Yuons to vote for Hun Sen, so Khmers should vote for Sam Rainsy to protect our territory.'

(Rainsy, cited in May, 2013: 1)

Attitudes of this kind, even amongst political leaders, are far from novel in Cambodia. Centuries of alternating and often burdensome vassalage to geographic neighbours, Vietnam and Thailand (Chandler, 2009), following the decline of the Angkor empire, have hardened a mindset of territorial vulnerability; of being surrounded by aggressors whose intent is to acquire yet more of the territory they have stolen for so long. Most prominently, 'Vietnamese are perceived by Khmers as having a long history of territorial invasion, since all of Southern Vietnam was once part of the Khmer kingdom' (Ninh, 2017: 219) and these long held cultural attitudes have been entrenched by official affirmation for many years.

Not only do 'maps showing the southern region of Vietnam as a part of Cambodia continue to be displayed at many public sites, including the Royal Palace and the National Library in Phnom Penh' (Ninh, 2017: 220), but anti-Vietnamese rhetoric a key part of political rhetoric since the first national elections in 1993 (Berman, 1996; Amer, 1994). In particular, an entrenched public discourse that Prime Minister Hun Sen and the Cambodian People's Party's historical links to the PRK administration make them 'Vietnamese stooges' who facilitate Phnom Penh's control by Hanoi (Hutt, 2016: 1), has rendered it a ubiquitous feature of opposition politics which 'transcends ideological differences' (Amer, 1994: 229).

During Cambodia's short democratic history opposition politicians from across the political spectrum have benefitted from stoking historical fears of invasion from the East (Lewis, 2015; Berman, 1996; Amer, 1994). However, since the CNRP's defeat in 'the rigged elections of 2013' (Lewis, 2015: 8), 'deep suspicion and hatred of the Vietnamese' (Frewer, 2016: 1) has manifested more frequently in action, with racial incidents, attacks, or even murder become 'fairly common' (Lewis, 2015: 8) in certain parts of the city. As some commentators have argued, this latest wave of prejudice—manifested not only in racist attacks and popular discourse, but also in rising police harassment (Parsons and Lawreniuk, 2018) and a sharp spike in deportations—is once again approaching a peak. Indeed:

Since April 2014, when the government established the general department of immigration, nationwide statistics show that 11,661 foreigners from 72 countries have been deported and permanently barred from the kingdom. Of them, 9,777 – or nearly 84 percent – were Vietnamese nationals. (Ros, 2017: 1)

However, this latest incarnation of anti-Vietnamese sentiment is a different prospect to its predecessors. Rooted in the same translocal and multi-scalar nationalism that has reinvigorated opposition politics in Cambodia, today's suspicion of the other is a mobile one, predicated not on static boundaries, but the same fluid systems of movement that have come increasingly to define this economically open and internationally integrated nation. Thus, just as the language and form of Cambodian nationalism—undergirded as it is by 'categories or principles of inclusion/exclusion in terms of the "we" and "them"' (Tesfahauney, 1998: 502)—have adapted to suit a more global and fluid mindset, so too has the 'closely related' (Tesfahauney, 1998: 502) phenomenon of prejudice.

Rather than being a peripheral feature of the battleground of inclusion, therefore, mobility—and documentation, which acts as shorthand for the rectitude of mobility—is increasingly at the centre of Cambodia's prejudicial narratives, acting a key means of distinction between the 'ins' and the 'outs' of nationalism. Specifically, the intertwinement of the labour movement with nationalist politics first became apparent during the national protests of 2013 and 2014, during which 'clear anti-Vietnamese sentiments emerged, as protestors harassed Vietnamese Cambodians, and the opposition accused Hun Sen of seeking Vietnamese assistance in defending his power' (CANVAS, 2016: 21) and would culminate a few months later in protests outside the Vietnamese embassy in July 2014, 'demanding Vietnamese recognition for Cambodia's historical claim of the Khmer Krom territory, transferred to Vietnam in 1949' (CANVAS, 2016: 22).

From the 2013 elections until the dissolution of the opposition in late 2017, the CPP and CNRP competed with each other to show strength in enforcing regulation of a notably porous and historically mobile Eastern border (Singh, 2014), as repeated opposition accusations 'of being under strong influence from Hanoi, of ceding territory to Vietnam and of allowing Vietnamese immigrants to illegally enter and work in Cambodia' (Vannarith, 2015: 1) were been met by ostentatious government action. Without prejudicing top level diplomatic relations between with Vietnam, which have generally remained cordial, data released by the police authorities in 2017 bears out the extent of the government's involvement in this battle for nationalist sentiment. The number of people deported to Vietnam more that quintupled from 1059 to 6265 between 2014 and 2015, according to statistics from the ministry of interior (Khy, 2017), with 2016 seeing a reduced but still well above average figure of 2453.

Nevertheless, though substantial, deportations account for only a small part of the enforced mobility engendered by nationalist narratives. Although the level of deportation has risen notably in recent years, the 5 per cent of Cambodians who are ethnically Vietnamese (Rumsby, 2015) are troubled to a far greater extent by the everyday intransigence of officials in relation to documentation. Ethnically Vietnamese inhabitants of Phnom Penh, reported that it is sufficient simply to appear Vietnamese in order to encounter obstacles to obtaining documentation, with one stating, for instance, that 'when I was called to do the ID card, the official looked at me from head to toe and said that I was Yuon' (ethnic Vietnamese woman28/07/2015). Moreover, such testimonies are neither isolated, nor the result of personal prejudices on the part of administrators, but the product of a systematic process of exclusion. As the head of an urban village in Phnom Penh explained:

> I can make a letter for them, but it will say 'Vietnamese nationality' on the letter. For those who rent land here, generally we don't give any letter to them at all. Even if they own their land, if they aren't Cambodian nationals, then we can't make anything, even a family book.
>
> If they are from Kampuchea Krom, then I can do it from them. The way of speaking is different, the accent is different and also the way they look is different. Also their colour: [people from] Kampuchea Krom are black, but Vietnamese [people] are white. And [people from] Kampuchea Krom, even if they stay here for a long time, then they still speak unclearly, whilst Vietnamese people, if they stay here for a long time, they can speak clearly.
>
> (Village Chief, Phnom Penh, 19/06/2017)

As such, the testimony of this official not only emphasizes the role of aesthetic judgements in decisions over whether or not to provide documentation, but also the geopolitical undergirding of such judgements, tied as they are in this case to nationalistic perceptions of Kampuchea Krom. In addition, despite their often individualised basis, official attitudes such as these are often extended to entire families or communities, regardless of their migratory history or the specifics of their heritage. As ethnically Vietnamese people born in Cambodia explained, 'if the authorities think that your parents are Vietnamese, then you can't get ID even if you were born in Cambodia' (ethnic Vietnamese worker, 19/05/2017). Similarly:

> Even though our parents are Vietnamese and they were born in Vietnam, we were not. Just like most people round here we were born in Cambodia so we are Cambodian, but we cannot get ID. [It's true that] I know of some relatives in Vietnam, of my parents' generation, but I don't even know if they're alive or dead now, as I've never visited them and they've never come here.
>
> (ethnic Vietnamese man, Chbar Ampov)

This inability to access documentation, even for those who have previously possessed it (Parsons and Lawreniuk, 2018) generates a raft of inequalities for Cambodia's ethnically Vietnamese population (see Figure 7.3). Firstly, the inability to register a birth impedes access to schooling, meaning that the objects of official prejudice 'stay illiterate... [and] ... can only work in jobs like construction, or others that require strength, so life will be difficult for them' (ethnic Vietnamese

Figure 7.3 Floating houses of a Vietnamese community in Northern Cambodia, 2018. Many ethnic Vietnamese in Cambodia live on boats or floating homes on the nation's waterways as a lack of identity documentation prevents them from owning land. *Source*: Author.

woman, Chbar Ampov). More immediately, however, the inaccessibility of documentation generates strong occupational restrictions, entrenching precarity in the livelihoods even of those who already possess documentation:

> If you go to work and the police catch you, then they will send you back to Vietnam. Even if you're born here and the police catch you, then they will fine you $200-300 and they don't release you until you pay.
>
> (ethnic Vietnamese man, 03/08/2015)

Furthermore, this uncertainty of livelihoods manifests not only in those who are caught, but far more broadly, as ethnically Vietnamese inhabitants and citizens of Cambodia are forced to rely on informal and insecure income. Indeed, as a welder who provides services to his local community explained, 'if you don't have money to set up a business, then it's very, very difficult' (ethnic Vietnamese man, 03/08/2015), leaving many with few options beyond 'just cleaning somebody's house or doing wage labour. You can't get any jobs' (ethnic Vietnamese manan, Chbar Ampov). Even for those lucky enough to secure employment, security of income is often minimal, with options heavily restricted by the reluctance of employers to hire people lacking the relevant documentation:

> For instance, my son in law is Vietnamese and only has a name in a family book, but he works as a construction worker. He feels scared [of arrest] but there's

nothing he can do, just keep making money and doing his job...It's difficult [to move between jobs]. They can only do wage labour: just $10 for a few days and then stop. Also the owners of some workplaces are scared to take people who don't have ID.   (ethnic Vietnamese woman, Chak Angkrey)

However, although such inequities exist throughout the labour market, forcing those without a complete set of documentation to participate in less formal patterns of labour than those who possess one, it is factory work—and in particular the garment industry—that excludes the partially, or non-documented population to the greatest extent. Oversubscribed to the extent that departing workers are replaced within a day on average (CARE International, 2017) and subject to Cambodia's only statutory minimum wage, the garment industry has little need to employ anybody who could potentially prove a problem to their business. Testimonies emphasizing the extent of this exclusion were repeated again and again by informants, who explained that 'we want to get a job in factory but we can't do it because we have no ID. None of us have jobs because of this' (ethnic Vietnamese construction worker, 03/08/2015); that 'they will not allow me to work in a factory because I don't have the papers' (ethnic Vietnamese woman, 19/06/2017); and that the impossibility of obtaining them amounted to a permanent exclusion from garment work. For example:

We had documents, but now they're all gone. Now no one will make them for us...I have a letter that proves I have been here for a long time, but not an official one...It doesn't change our plans because we don't work [formally] for anybody. I am Vietnamese, so if I go to work in the factory, then they won't accept me.
(Focus group, ethnic Vietnamese women, 15/06/2017)

This de facto disbarment from the largest formal employer in the country (as well as the more formal employers within the construction industry), has effectively enforced an informal livelihood on much of the ethnically Vietnamese population, a state of affairs internalized on both the Khmer and Vietnamese sides. Officials view these informal livelihoods as matter of choice, arguing that 'generally, the Vietnamese don't work in the factory, so they don't need any help from me, they usually just plant morning glory' (Village Chief, Phnom Penh, 19/06/2017) and informal workers highlight the benefits of their livelihoods in the context of low income and insecurity, explaining that 'if we send our children to work in the coffee shop, then we can borrow money from the owner [against their salary] one or two months beforehand' (focus group, ethnic Vietnamese women, 15/06/2017). Furthermore, as a daily wage labourer elaborated:

People here prefer to do their own business. Only Khmer people go to work in the factory. We think if we can do our own business then we can get money every day. In the factory, the salary is monthly, but children need to eat every day.

(ethnic Vietnamese labourer 15/06/2017)

Taking a broader view, however, the centrality of the garment industry to Cambodia's economy and politics alike highlights the translocal character of prejudice and exclusion in the Kingdom. The same new politics that has utilised the multi-sited nature of factory-based livelihoods to draw together diverse occupational sectors and bridge rural–urban rifts that have dominated the national mindset for decades (Davies, 2010) if not hundreds (Chandler, 2009) of years has, at the same time, entrenched exclusion at the margins of society. Cambodian unity has become *Khmer* unity, as an increasingly mobile workforce turns to ethnicity rather than geography to demarcate national inclusion. Ultimately, it is this very hardening of moods, manifested through the politically malleable mechanism of documentation, that has served most effectively to immobilize the excluded in the trap of informality.

Thus, in sharp contrast to the emphasis within the language of Cambodian nationalism on insiders versus outsiders; the static rectitude of the Khmer people set against the aggressive outsiders who must be repelled from the communal home, Cambodia's age of translocality has seen the battlefield of nationalist discourse transition to the realm of mobility: those who are in, move; those who are out, stay in place. The politics of translocality is therefore one in which mobility is both definitional and punitive. The threat of deportation and the inability to access documentation combine to enforce mono-locality upon outsiders, who are left static by the movement of others. In this context, mobility becomes an inherently political process, feeding into and drawing from a nationalist narrative that is moulded by the geographical complexities of contemporary livelihoods. The nexus of mobility and identity, always intertwined, has itself moved to the heart of national politics.

Conclusion

Though their mobility has sometimes been disguised by the enduring myth of intransigence that pervades external impressions of the Kingdom (Chandler, 2009), 'Cambodians have long been on the move' (Edwards, 2007: 421). On a grand scale, indeed, Cambodian identity is in many ways defined by a historical readiness to migrate, with transient borders and the loss or gain of territory providing a sense of vulnerability that pervades the character of the nation. Yet this reality has been 'subsumed in colonial discourses on Cambodia' (Edwards, 2007: 422) which continue to lie at the core of the production of nationalism. Mobility has become highly politicised and interlaced with normative judgments, as debate over the legitimacy of one type of movement or another move increasingly to the centre of nationalist discourse.

Indeed, although historical trends are instructive, neither mobility itself nor its influence on politics is constant. Key developments both in the nature of work and the technology popularized by shifting livelihoods have brought about concomitant changes in the mindset of workers and nations alike. In particular, the influence of mobile telephony combined with improvements to infrastructure and transport have allowed rural–urban flows of people, money and goods to adopt an increasingly central place in the Kingdom's economy. Above and beyond the ebb and flow of the political landscape, it is this new, multi-sited conception of place that lies at the core of the 'ultra-nationalism [that] has quietly colonized emerging opposition to the current regime' (Frewer, 2016: 1). Domestic mobility is increasingly a shibboleth of legitimacy, whilst international mobility—even that undertaken generations previously—has become a narrative trope of exclusion, denoting a dislocated or bifurcated self.

In its characteristic notes of prejudice—accusations of historical theft and ethnically mediated discrimination—this new translocal nationalism shares much not only with its predecessors, but also the rising tide of nationalism globally. However, the aim of this chapter has been to capture something of its distinction from these wider forces. By elucidating the influence of Cambodia's contemporary, specific and everyday systems of movement on inclusion and exclusion, it has asserted not only that 'workers' politics cannot be expressed in state–civil society relations', but also that it must be attuned to the 'repoliticisation of labour' underway as national and industry agendas converge (Arnold, 2017: 23). Within an economy increasingly defined by accelerating patterns of movement, the politics of exclusion has therefore transitioned to a politics of (im) mobility.

8

Framing a Total Social Fact

Inequality, famously, is a slippery and evasive beast. The phenomenon has been described, in various contexts, as an 'ogre' (The Guardian, 2018: 1), a 'three headed hydra' (The Economist, 2014: 1) and a 'seven headed dragon' (Van Den Brink and Benschop, 2012: 71), reflecting a phenomenon with 'a multitude of faces in different social contexts' (Van den Brink and Benschop, 2012 71). It is at once poise and poverty; passed exams and flooded fields, but—to the frustration of those who recognize its looming presence and seek to curb it—it is none of these things, nor even a definable set of them. Individual measures may reveal one 'face' of inequality, but in doing so, they reveal little or nothing of its totality. The challenge lies not in identifying and measuring inequality within pre-set categories, but linking them; understanding how each emerges from and feeds back into the whole.

Nevertheless, to do so is far from straightforward. As with the most accomplished mythical foes, the beast hides in plain sight, shrouded by doubts over its existence and importance. Economists in particular, have not 'expressed much concern about widening inequality' (Pickett and Wilkinson, 2009 x), whilst various economic frameworks have offered comforting axioms in this regard. Kuznets' (1955) famous inverted U, for instance, explains that wealth inequality is not only a sign of a healthily growing economy, but merely a transitory state preceding a sustained return to the mean. More recently, Milanovic (2016) has posited an undulating historical model of income inequality dating back to the thirteenth century, casting the rise and fall of global inequality as a slow moving historical process; 'a necessary, if not inevitable, consequence of a growing economy' (Pickett and Wilkinson, 2009 x) and one far removed from the agency of governments, firms, and individuals.

Part of the reason for this lack of concern is that, even if inequality is permanent, disadvantage itself is viewed as transient. Those who are poorer now need not always be the ones who are poorer. It is enough for the models merely that—for reasons of the competitive incentive—*somebody* is. However, as noted to widespread acclaim by Thomas Piketty (2014), such assumptions rely overwhelmingly on income measurements, an imperfect approximation of wealth that elides the historical advantages hidden in capital. Indeed, in recognizing that wealth is more complex than either income or money, economists are beginning to appreciate what researchers from Victorian Britain to the contemporary developing world have long since known: that inequality, like the absolute condition of poverty, is

Going Nowhere Fast: Mobile Inequality in the Age of Translocality. Sabina Lawreniuk and Laurie Parsons,
Oxford University Press (2020). © Sabina Lawreniuk and Laurie Parsons.
DOI: 10.1093/oso/9780198859505.001.0001

multi-dimensional, complex, and self-reinforcing; an entity beyond its discrete manifestations.

However, it is in reality more fundamental than this. From the food, low on both volume and quality, that stunts development and shortens lifespans, to the mannerisms, habits, and preferences that stratify rich from poor by giving poverty the appearance of nature's work, inequality is so deeply embodied as to become invisible under ordinary circumstances. Often, it is only in times of change that 'like an enormous rock in a creek' (Galtung, 1969: 172) it begins to form the currents and eddies that reveal its presence. These disruptions to the steady flow of economic progress have been the context of this book, yet the rock itself is the subject. By focusing on mobility, it has aimed to reveal inequality's hidden, multi-faceted permanence, recasting it from ephemeral entity to 'total social fact'. This chapter will attempt to draw together these lessons, in pursuit of a framework and method suitable for the discernment of a phenomenon whose permanence is best revealed by the mobility of its environment.

8.1 Framing a Total Social Fact

The concept of a total social fact is one well known to sociologists. Emerging from the idea of social facts, viewed by Emile Durkheim (1895) as the basic building blocks of society, it is a nodal principle from which others derive meaning, providing ballast and structural integrity in a dynamic environment. Thus, 'a total social fact is such that it informs and organizes seemingly quite distinct practices and institutions' (Edgar, 2002, 157). It is both a point of entry and an object of interrogation in itself, but it is fundamentally a point of immobility, a socially styled pole of objectivity around which more transient and subjective norms revolve.

This book has made use of this idea as a means to reconceptualize inequality. However, in doing so, we offer a distinct interpretation. Where we refer to the total social fact of inequality, we view its permanence and inurnment to the multiplicities of translocality not as the result of immobile consensus, but as derived from its very mutability. As the post-war planners who built the British comprehensive schools referred to in Chapter 1 found, inequality is endlessly protean. No sooner has one part of the playing field been levelled than do divots and hillocks appear elsewhere in the landscape of wealth. It is a phenomenon apparently freed from the normal rules of structure, occupying all spaces and scales with equal ease and expending no effort to traverse them. Nevertheless, despite its weightless mobility, inequality viewed as a total social fact is readily observable. It exists in the most physical of processes, steering action and move-ment according to highly structured patterns of disadvantage.

The implications of this position for physical mobility are substantial. As interest in the multiple meanings of movement has grown, an assumption of scalar compatibility has nevertheless been retained: movements are not supposed to have implications far beyond and outside of their own manifestation. Yet, as the various contexts and personal accounts included here have sought to demonstrate, this is exactly what occurs. National narratives of disadvantage drive household scale mobilities and vice versa. However, thus far the inherently static epistemology characteristic of migration studies—wherein mobility is set up as the corollary to the norm of immobility—negates a fundamental truth about mobility: that it never ceases, but merely jumps scales and shifts in form; that, immobility, otherwise put, is only ever a partial state, frequently impelled by mobility in other spheres.

Viewed thus, economic development—especially on a global scale—raises a key question: if mobility has increased in some ways and for some people, where has the immobility associated with disadvantage manifested? Answering this requires abandoning the idea of immobility as a starting point. Indeed, despite 'a wide-spread historical tendency to view indigenous societies ... as historically emplaced: that is, as the product of a long history of engagement with particular locales' (Alexiades, 2013: 1), growing evidence suggests that it is in fact complex, multi-sited and multi-faceted patterns of movement that constitute the global and historical norm. Yet, in the global North and South alike alike, the 'myth of the immobile peasant' (Skeldon, 1997: 7–8) remains a highly influential discourse. Migration, development, and social change are all coloured by the notion of having begun, at some more or less recent point, from zero.

In Cambodia, this narrative of stasis acquires an especially stark relevance, given its historically subjugating role. The perception of the Kingdom as a happily anachronistic nation, living 'day by day, poor but contented' (Poorée and Maspero [1938] in Derks, 2008: 30) fitted perfectly with the ethos of those agents of the French *Mission Civilitriste* 'who in the nineteenth and twentieth centuries saw themselves as introducing change and civilisation to the region' (Chandler, 2000: 10). Whilst the French did not create this caricature of immobility, which has its roots in the 'deep continuities [and] refusals' of Cambodian culture (Chandler, 2000: 247), the maxim of immobility has been 'manipulated by different political players throughout history, creating a situation where nationalism is imposed vertically from the powerful elites to the general Khmer population' (Grant, 2009: 30).

Moreover, as the accounts in this book show, the nexus of mobility and identity is not only alive, but thriving in contemporary Cambodia. Political leaders, fuelled by the dynamism of their industrial grassroots support, have seized once again upon mobility as a badge of inclusion and progress. Those who move like us, they suggest, are with us, whilst those who move otherwise—or not at all—are against us. Mobile inclusion therefore represents dynamism and progress, but also more conceptually static categories. The ubiquitous national appellation employed by politicians, 'srok Khmer yeung' [our home of the Khmer people], includes at least three in itself; bounding, legitimising, and excluding in a single verbal turn. To be an outsider, according to this rhetorical logic, is to be inherently an imposter; one who moves without justification into a place they should not be.

Nevertheless, whether or not they are employed knowingly by those who seek to benefit from them, exclusionary discourses of mobility such as these are not the creations of those who purvey them. Rather, manipulated, deployed and dressed in the colours of a cause though they may be, they are the product of—and indeed, an integral part of—existing structural inequalities, rather than an extraneous addition, or catalysing factor. Where Cambodia's leaders and would-be-leaders contrast inclusion and exclusion, therefore, they are not dividing anything as such, but merely applying a pre-existing narrative to a pre-existing mobility differential in order to accentuate and articulate it. Mobility creates boundaries of its own accord. The only question is where they fall.

Away from the dominant voices of key political actors, this point is perhaps best exemplified by prejudice at a far smaller scale. In particular, the evocative narratives surrounding the rectitude of begging behavior highlight that no leading voice is necessary for narratives of exclusion and inclusion to become bound to patterns of mobility. Rather, the denigration of Cambodia's mobile alms seekers is formed agglomeratively, coagulating from the raw material of folklore and tradition. It is therefore not created, but evolves in multiple places simultaneously, as an articulated expression of structural knowledge between people who share the same space, but whose mobilities are distinct.

The myths surrounding beggars are instructive here. The truth is not that people are inherently suspicious of mobility—it is, after all, the less physically mobile ethnic Vietnamese who are vilified in national discourse—but that suspicion arises where perceptions of wealth differ from familiar patterns of mobility. Few complaints are heard when those who find themselves in difficulties resort to begging in their own locality; many beggars report receiving charity from their neighbours in times of need and systems of indefinite, interest free lending are a common means by which rural Cambodians smooth consumption (Parsons, 2017). Such behavior becomes objectionable only when it is paired to patterns of mobility with which it is not usually associated.

Mobility, otherwise put, has a normative character derived from prior systems of movement. Since those who most commonly engage in high levels of physical mobility have most commonly been garment and construction workers engaged in waged or salaried labour, a particular socio-economic profile is associated with their patterns of migration. Where people whose circumstances differ from this engage in the same or similar patterns, explanatory narratives begin to emerge and diverge. The fierce battle of discourse raging in Prey Veng province, between those who claim a socio-economic position—destitution—that is a normatively acceptable basis for movement in the eyes of their community, and those who seek to deny them this justification, exemplifies this process in real time: since mobility requires a rationale to explain and legitimate it, the price of mobility is normative change.

This change, though, need not be negative. As outlined in Chapter 4 and by various authors on Southeast Asia (e.g. Lawreniuk and Parsons, 2017; Parsons and Lawreniuk, 2014; Derks, 2008; Conradson and Mackay, 2007; Elmhirst, 2008 translocal systems of movement can engender narratives of praise as well as denigration. Women, in particular, have experienced a change in the interpretation of their household roles in recent years, from a norm of relatively low mobility to one in which female migration – both domestic and international – has become not only commonplace but economically vital to many of the communities in which it has proliferated.

As with the narratives of denigration outlined above, this has involved a process of normative contestation, as long distance movement conflicts with both informal norms and the formal codification of women's roles set out in the *Chbap Srey* [women's law]. In this case, however, normative re-articulation has eased the disjuncture. Where women of marriageable age were previously expected to *go into the shade [joel m'lup]* by spending up to six months inside their homes to lighten their skin, the same norm is today fulfilled in the shade of modern sector labour (Derks, 2008). The role of familial supporter and caregiver, similarly, has metamorphosed from a domestic setting to a wider, marketizing context, so that, as the village chief of a high migration village asserted, 'now it is better to have daughters than sons because daughters help the family' (Village Chief, Prey Veng, cited in Parsons et al., 2014).

Changes such as these have re-imagined the language, if not the meaning, of gender roles in the developing world. However, it has been a persistent failing of migration studies to assume that normative change stops here, confined to economic agents and their immediate environs. In reality, the ripple effects of movement: 'mobility, connection, emotion' (Conradson and McKay, 2007: 167) spread far wider, so that 'immobile actors are often beholden to the same translocal subjectivities experienced by their mobile counterparts' (Platt et al., 2016). Mobility, in other words, transforms the life-worlds of the immobile just as readily as it does those of the mobile. As demonstrated in Chapter 4, supposedly non-wage earning members of translocal households—especially those in caring roles such as mothers, siblings and grandparents—play an active and vital role in migration systems, adapting their mobilities and lifeworlds in complex ways to the changing demands they face.

Nevertheless, the generalized excision of older people and those in caring roles from the focus of migration studies demonstrates the persistent scalar and economic bias in the treatment of mobility. Even those analyses purportedly focused on the social and emotional dimensions of movement tend to position themselves, implicitly, at the scale of economic flows, thereby diminishing the importance of the mobility underway at either a different scalar footing or enacted for ostensibly non-economic reasons. The mobilities associated with visitations, childcare, and the adoption of a leading role in the household have consequently received far less attention than those which results in a wage. Yet, it is precisely these less tangible actions and reactions that are the basis of inequality, driving normative change and advantage in a manner that economics often fails to capture.

Indeed, what is key here is not only that pastoral roles are useful in supporting an income generating migration system, but that in doing so they validate and protect the status of those who perform them. The many Cambodian grand-mothers who criss-cross the country to maintain linkages between migrant chil-dren and their offspring—like those who adopt responsibility for the family farm—are at pains to emphasize the value of their work, highlighting the praise they receive, the voluntary adoption of their new roles and the economic benefits it brings to their families. In doing so, therefore, they resist the marginality that threatens those unable to access modern sector wages in a marketizing context, demonstrating not only the capability to effect normative change, but also an awareness of the imperative to do so.

Furthermore, by protecting their status in this way, Cambodian caregivers demonstrate, implicitly, what is missed by those who cannot do likewise. Although far from a zero sum game, changes in status in one sphere usually produce impacts elsewhere and the increasing prominence of factory work across the developing world has left those unable to access it at a disadvantage. In particular, the garment industry's gender bias—women are both more likely both to apply and be accepted to undertake work in the garment industry's entry level positions (ILO, 2012)—has meant a reduction in male status in migrant sending communities that has contributed to rising delinquency and gangsterism amongst young men railing against the growing ignominy of their mobility (e.g. Czymonewiecz-Klippel, 2013; Elmhirst, 2008).

What cases such as these serve to highlight is that the interplay of economic and normative disadvantage, though close knit, is complex. Viewing inequality as a total social fact means re-positioning, expanding and adjusting existing lenses to account for such complexities. Nevertheless, in emphasizing the mutability of disadvantage, it does not mean drawing the study of inequality into a nebulous, or theoretical realm. Rather, the opposite is true. As identified in the testimonies and data herein, inequality is a physical state, inscribed on bodies and furrowed in the land. What is key is to recognize the interaction of abstract and often ephemeral processes—narrative, discourse, denigration, adulation—with permanent, tan-gible, and highly physical forms of disadvantage.

For this reason, the account of inequality presented herein has sought as far as spossible to maintain a connection to the land and its physical employment. As shown throughout this book, small differences in natural assets are key to understanding the genus and perpetuation of inequality in mobile systems. In a changing climate, these are often incidental, as the value and utility of a piece of land or natural feature may have shifted unexpectedly in the context of unpredictable rainfall and rising temperatures (Oeur et al., 2012; MoE and BBC, 2011). It is the highly mobile socio-economic environment that sees these changes locked in by the structural and narrative mechanisms that transmute transient capitals into long-term wealth.

Nevertheless, locking in does not mean tying down. Rather, what this book and in particular Chapters 4 and 5 have sought overarchingly to demonstrate, is that this process of entrenching the advantage and disadvantage conferred by the land does not necessarily mean attaching it either to a particular place or form of capital. Once passed from the land to the people who depend on it, it is freed from its spatial moorings and may manifest anywhere within the translocal system it forms part of. The consequence of this is 'telecoupled' (Baird and Fox, 2015: 4) ecologies in which migration systems connect, re-use and degrade otherwise discrete agro-ecological zones, thereby linking land use change into the same global political economic structures as the migratory flows that catalyse it.

Viewed thus, the large scale processes of ecological change that have taken place in Cambodia in recent years in fact tell only part of the story. Alongside rubber plantations, deforestation, and mining concessions, the Kingdom has borne witness to a subtler but also more widespread agro-ecological shift. Mechanization and the marketization of agricultural inputs, rising rural wages driven by modern sector opportunities and an increasingly capricious climate have driven widespread but thus far little noted, shifts in rural practice. Capital intensive broadcasting is increasingly replacing more labour-intensive transplanting practices, inducing smallholders to depend on modern sector remittances to farm. However, not everyone can depend upon wage earning relatives. Those unable to access these funds have two choices: sell or rent to larger landowners, or take on debt to pay for inputs; an option that often leads ultimately to sale, as crop failures leave farmers unable to repay loans. In either case, the ultimate result is land concentration in fewer and fewer hands and growing pressures on NTFPs, fisheries and migrant receiving ecologies.

That over 40% of the country is now landless or near landless (ADB, 2014)—especially coming from a position of zero landlessness in the 1980s following land redistribution by the PRK government—is testament to the speed with which this process has occurred. However, despite its scale, it is a phenomenon that has rarely been explored holistically, viewed instead as the aggregation of individual experiences which share only their guidance by the invisible hand of the market. Consequently, little has been understood of the human inter-linkages that structure this national process of repurposing; the multi-local impact of degrading ecologies and the manner in which global geopolitics leave their mark on the Cambodian soil.

Indeed, what our account has aimed to show is that ecology is caught up in the same multi-scalar narratives of work, merit and status as are other dimensions of mobility: rice land becomes a site for cash crops because of 'dutiful daughters [and] broken women' (Derks, 2008: 170); transplanters become broadcasters because of 'rice people in the city' (Derks, 2008: 21) and smallholders become landless because, as the proverb goes, 'drip, drip, the vessel fills [whilst] pouring causes it all to spill' (village focus group 03/06/2016). Viewed thus, stories are the bedrock on which translocal decisions are made. As part of the driving equation of mobility, these shared narratives help to replicate and transport ecological change across multiple sites, linking them inextricably into multi-scalar political-economic structures.

The mutual intelligibility of these narratives—from a shared approval of debt for investment, borne in large part of the recent boom in microcredit (Bylander, 2015), to the widespread practice of purchasing land from those suffering repeated agro-ecological pressures (Parsons, 2017)—ensures that much of the ecological change most crucial to the lives of ordinary Cambodians is fundamentally inter-twined with discourse. Floods and droughts do not therefore cause land sales in isolation, but are merely the acute dimension of a systematic set of pressures—ecological, economic and narrative—that have progressively deepened the pre-carity of smallholder livelihoods, stripping away the conceptual safety nets entreating the rich to protect the poor 'like the cloth that surrounds' them (Fisher-Nguyen, 1994: 99) in favour of a new set of market-driven imperatives.

In Cambodia, this market-environment-normative nexus is central to liveli-hoods around the country. Yet whilst certain patterns of normative change are familiar from previous studies on agrarian transition (e.g. Platteau and Abraham, 2002; Fafchamps, 1992; Scott, 1972)—a greater social distance between richer and poorer people; a lack of interest in reciprocity on the part of the wealthy—these have yet to reprised in the light of translocal livelihoods. As the influx of modern sector remittances has changed the character of labour and information sharing, this book has aimed to show that these processes of normative shifting are as relevant as ever.

Indeed, rather than the safety net they once constituted, these traditional systems of reciprocity and mutual support increasingly perform a socially strati-fying role, encouraging partnerships between machine users and large scale renters on the one hand and smallholding retainers of traditional methods on the other. The norms which once provided a basic level of security against the vagaries of agriculture have therefore now been transmuted in character, serving to limit rather than guarantee the livelihoods of poorer Cambodians. The result is that the interplay of communities with the land on which they depend—never in fact a lottery—is increasingly a game they cannot win.

Generating a perspective capable of conceptualizing these structures of pro-gressive disadvantage within the wider context of economic and ecological change is a clear objective in principle. Nevertheless, as detailed throughout this book, the challenges associated with such an approach are manifold. Norms are not easily quantified, for instance, and multi-scalar systems present significant difficulties to model, especially when feedbacks are given due consideration (Porst and Sakdapolrak, 2017). Stories, relative and inherently hermeneutic, sit uncomfortably with the atomistic and ostensibly objective statistics, presenting a range of difficulties to the establishment of a transferable method. Yet an alternative is viable: the establishment of a core gestalt of methodological exemplars whose principles and approaches may serve as a means to cohere an otherwise variegated body of literature on translocal inequality.

This will necessitate a broad set of tools. Indeed, the variety of approaches employed in this book alone bears testament to the diverse methods required to broach the investigation of economy, ecology, and discourse. Social network mapping, quantitative surveying, statistical analysis, qualitative interviews, and focus groups have each played a vital role in data collection, but this should by no means constitute an exhaustive methodological array. Rather, the strategies used here should serve first as an example of how diverse or novel methods may be utilized to measure inequality, but also secondly as a guide to the propositions that underpin their invention and deployment.

Broadly summarized, two principles predominate. First, mixed methods are not only advantageous, but essential in studying both mobility and inequality. Both the numerical and normative dimensions of inequality are partial in isolation, with neither able to convincingly explain inequality's persistence in a mobile environment without reference to the other. Numerous examples have been given in previous pages to demonstrate this, but the beggars of Prey Veng province provide an especially instructive case of this dual disadvantage: a lack of physical resources drives their migration—often enhanced by acute shocks or long term climate pressures—but it is an articulated norm, rather than poverty, that underpins their denigration. Classified as breaching moral codes in both rural and urban areas, their temporary disadvantage therefore becomes absolute deprivation, yet without the combination of quantitative and qualitative data this crucial distinction may be missed.

Understanding differences in the inequalities faced by older people, women and men requires a similarly dualistic lens. Norms of gender and age define the boundaries not only of the possible, but also the probable, creating differential structures that shape movement and occupation. Older people, for instance, know that they have little to no chance of securing modern sector work, making the absence of a productive family role a source of denigration and often reduced financial and emotional support (see also Lawreniuk and Parsons, 2017). Conversely, norms of care faced by female migrants—and in Cambodia, especially the youngest children in a household—serve in many cases to lock those who are able to secure urban work into long-term working penury, remitting all but their most basic needs for a potentially indefinite period. In both examples, therefore, economic factors are the dependent variable in a normative structure that disadvantages the object. Information, in other words 'is always the combination of data and meaning' (Schaffer, 2017: 8).

To move beyond a merely contextual perspective, however, the use of mixed methods is a necessary but insufficient condition. If the analysis of inequality requires a multi-scalar lens, it also requires a multi-scalar methodological toolbox, yet convincing efforts to compile one remain limited. Indeed, whilst 'social science – and migration studies in particular – have decidedly moved away from spatially-bound frameworks since the 1990s' (Xiang, 2013: 282), this focal expansion has been conducted overwhelmingly on a single scale. Multi-sited studies are now common in the analysis of mobility, but the scalar dimensions of movement remain underplayed to an even greater extent. Economic, ecological, and normative inequalities all have their origins at multiple scales, so that their concurrent examination 'enables an explanation of why some mobility is more consequential than others, and identifies strategic sites where critical engagement can be grounded' (Xiang, 2013: 282).

These lacunae drive to the heart of the current, cross-cutting academic interest in inequality: the recognition, in other words, that our tools thus far have captured only a small part—perhaps merely the symptoms—of the phenomenon. Yet acknowledging the need for a perspective on inequality capable of capturing both its translocal and multi-scalar dimensions has disciplinary implications also. Geography, in particular, with its history of multi-scalar (e.g. Swyngedouw, 2004) and translocal (Greiner and Sakdapolrak, 2013; Brickell and Datta, 2011b) scholarship, appears well placed to lead such a reformulation. Yet it is Development Studies to which the most significant contribution of such a framework might accrue, offering as it does a springboard from which to redirect the discipline towards its own classical traditions of 'redistribution and progressive forms of transformation in the context of stark asymmetries of wealth and power' (Fischer, 2019: 426).

Specifically, the book offers three contributions both to Development Studies and the study of inequality and mobility more broadly. First, it demonstrates not only the benefit, but rather the imperative of examining inequality in a translocal manner, incorporating linked, multi-site data collection and a degree of sensitivity to the place based interconnectedness that shapes obligation and opportunity. Second, the book aims to encourage a shift in the binary 'origin and destination' conception of migration which continues to influence much work in the discipline, towards a more dynamic conception in which multiple sites are dynamically connected through networks of communication, resources and affection.

Finally, combining both of these aims, the book has sought to make a dual headed case for a fundamentally mobile approach to the study of inequality in development. As shown here, not only is mobility dynamically shaped by forces operating in multiple places and at multiple scales, but so too are the social implications of that mobility. The interlinked, multi-sited perspective on inequality outlined here therefore provides a novel lens through which to examine the intersection of economy, ecology, politics, and culture and more broadly to examine the key issues under examination in development studies.

More specifically, moreover, it aims to resolve something of the tension between advocates of a more critically engaged and globally interlinked perspective on inequality in development (Horner and Hulme, 2019) and those seeking a return to a more transformative model of development evident in earlier development paradigms (e.g. Fischer, 2019). As argued here, the interconnectedness and rejection of North–South binaries advocated by Horner and Hulme (2019) amongst others need not be incompatible with the more active redistributive aims encouraged by Fischer (2019), for example. Indeed, as we have aimed to show, inequality cannot be ameliorated without fully understanding the complex means by which it resists and persists even in the most dynamic of circumstances.

Rather than offering a framework in the technical sense, the studies herein constitute a heuristic gestalt with which to demonstrate the value of a mobile lens on inequality. By combining data collection enacted on more than one scalar footing with qualitative accounts of how norms manifest across multiple sites, this set of interlinked perspectives therefore demonstrates how narratives of social inequality are rooted in and shaped by multi-scalar translocal livelihoods. Viewed thus, gender roles prescribed to a nation by the *Chbap Srey [women's law]* are reinterpreted in response to global economic processes and shifting household practices; older people resist global narratives of obsolescence with locally engendered normative change; and migrant beggars' patterns of mobility are driven both by national and local interpretations of the myth of an ancient curse. The complex interaction of global, international, and local inequalities, is revealed precisely such details.

From the perspective of development policymakers and practitioners, therefore, the Cambodian context provides ample exemplification of the need to apply a multi-scalar perspective to movement, both in terms of its meanings and the unequal determinants that drive it. Even beyond this, however, certain topics provide a particularly timely perspective on the complex inter-linkages of cultural processes with the global economy and environmental change. In particular, the mobile nationalism associated with Cambodia's garment industry—outlined in Chapter 7—demonstrates how mobility reshapes the scalar conditions that surrounds it, feeding back dynamically into the ecology, economy, and norms according to which it operates. This is therefore a nationalism that is shaped by translocality at multiple scales, with global economic forces tightly embedded in household relations, which themselves operate as key factors in mobility and the genus of protest. In an era in which not only inequality, but also nationalist politics and environmental degradation are the order of the day, there are, we believe, timely lessons to be learned from this microcosm.

Moreover, these multi-scalar interconnections are not a one-way street. Garment workers' sense of disadvantage is not passively engendered by local and global factors but exists in active contestation with them. The impact of their collective action, both in terms of achieving rapidly rising wages in a key export market and demonstrating to workers elsewhere the potential efficacy of unionism and protest, has returned their agency to the global economy, shifting Cambodia's position within the global industry (World Bank, 2018). Indeed, if the World Bank is to be believed, the growth of the industry has slowed as a result of the higher cost of labour, a global scale outcome predicated at least in part on the local narratives of solidarity, protest, and fairness that have fuelled the success of the union movement. Resistance, in other words, is still capable of leaving a global mark.

Despite the breadth of its framing, though, we argue that the strength of this perspective on inequality lies in its grounding in everyday, physical concerns. The dynamic co-constitution of economy, ecology and discourse ensures that disadvantage in one sphere moves fungibly into another: the small size of a family plot of land may impact most egregiously on one family member only (perhaps the older or younger examples referred to above); or alternatively may manifest, as in the examples of beggars and ethnic Vietnamese, in the form of a denigratory narrative that becomes attached to whole communities. To understand inequality, therefore, means understanding the relationship of physical and discursive factors at the smallest and largest scales. No two lenses are too distinct in size, or broadly spaced to add insight to the analysis of stigma.

Rather than annulling the researcher's obligation to discern structure, this perspective necessitates a redoubling of analytical effort. If inequality is truly multi-scalar and translocal, then any interpretation derived from a single context can be no more than microcosmic. A keen awareness of the wider narratives and structures within which any given case resides becomes essential, as does the assessment of onwards implications. Even the smallest and most specific narratives of prejudice have significant impacts elsewhere and given the mutually constitutive nature of exclusion explored herein, these constitute an integral, rather than peripheral part of the subject. Framing a total social fact requires total social research. Only through a genuinely unbounded ethnography can mobile inequality may be accurately or appropriately discerned.

In this regard, we stand this research up against a discourse of labour mobility that is intertwined inexplicably with freedom; in which to marketize is to liberalize and to liberalize means unshackling economic flows from the social constraints that bind them. Human migration, from this narrow perspective, appears merely to be one component within this wider process: labour flows naturally according to factor price differentials—allowing for the presence of market imperfections—

and growth is the result. Yet the examples presented here make the case for an opposite narrative. As this collection of studies shows, it is the imperfections, rather than the flows, that are viewed as natural by those involved. Information asymmetries, patronage and prejudice are not externally imposed obstacles, but key components of the implicit, protean, yet durable structures that mould a society on the move. Inequality in the age of translocality is therefore the ghost in the machine; the undergirding logic that co-ordinates and directs the flow of resources in motion; that most visible of invisible hands.

References

Abebe, T. (2008). Earning a living on the margins: Begging, street work and the socio-spatial experiences of children in Addis Ababa. *Geografiska Annaler: Series B, Human Geography*, 90, 271–84.

Abebe, T. (2009). Begging as a livelihood pathway of street children in Addis Ababa. *Forum for Development Studies*, 36, 275–300.

Abreu, A. J. G. (2012). *Migration and Development in Contemporary Guinea-Bisseau: A Political Economy Approach*. Ph.D. Thesis. SOAS, University of London. Accessed 25/10/2017 at http://eprints.soas.ac.uk/14243/

Adger, W. N. (2000). Social and ecological resilience: are they related? *Progress in Human geography*, 24(3), 347–64.

Ahmed, F. E. (2004). The rise of the Bangladesh garment industry: Globalization, women workers, and voice. *NWSA Journal*, 16(2), 34–45.

Alexiades, M. N. (ed.). (2009). *Mobility and Migration in Indigenous Amazonia: Contemporary Ethnoecological Perspectives*. Oxford: Berghahn Books.

Amakawa, N. (2007) (ed.). *The Transition to Market Economies and Industrialization of the CLMV Countries*. Singapore: IDE-JETRO and NUS Press Singapore.

Amer, R. (1994). The ethnic Vietnamese in Cambodia: a minority at risk? *Contemporary Southeast Asia*, 16(2), 210–38.

Anderson, B. (1983). *Imagined Communities: Reflections on the Origin and Spread of Nationalism*. London: Verso Books.

Arnold, D. (2013a). Better work or 'ethical fix'? Lessons from Cambodia's apparel industry. *Global Labour Column*, 155.

Arnold, D. (2013b). *Workers' Agency and Re-Working Power Relations in Cambodia's Garment Industry* (No. ctg-2013-24). BWPI, The University of Manchester.

Arnold, D. (2017). Civil society, political society and politics of disorder in Cambodia. *Political Geography*, 60, 23–33.

Asia Development Bank. (2007). *Kingdom of Cambodia: Community Self-reliance and Flood Risk Reduction*. Technical Assistance Consultant's Report. Bangkok: Ministry of Water Resources and Technology.

Asian Development Bank [ADB] (2014). *Cambodia: Country Poverty Analysis 2014*. Manila: ADB Press.

Baird, I. G., and Fox, J. (2015). How land concessions affect places elsewhere: Telecoupling, political ecology, and large-scale plantations in Southern Laos and Northeastern Cambodia. *Land*, 4(2), 436–53.

Bamisaiye, A. (1974). Begging in Ibadan, Southern Nigeria. Human Organization, 33, 197–202.

Banerjee, S. B. (2000). Whose land is it anyway? National interest, indigenous stakeholders, and colonial discourses: The case of the Jabiluka uranium mine. *Organization and Environment*, 13(1), 3–38.

Bansok, R., Chhun, C., and Phirun, N. (2011). *Agricultural Development and Climate Change: The Case of Cambodia*. Phnom Penh: CDRI.

Barney, K. (2012). Land, livelihoods, and remittances: A political ecology of youth out-migration across the Lao–Thai Mekong border. *Critical Asian Studies*, 44(1), 57–83.

Bateman, M., and Chang, H. J. (2012). Microfinance and the illusion of development: from hubris to nemesis in thirty years. *World Economic Review*, 1(2012), 13–36.

Bateman, M., Natarajan, N., Brickell, K., and Parsons, L. (2019). Descending into debt in Cambodia. *Made in China Quarterly*, 23/04.

Baviskar, A. (2001). Written on the body, written on the land: violence and environmental struggles in central India. In Peluso, N. and Watts, M. *Violent Environments*. Ithaca: Cornell University Press, 354–79.

Beresford, M., Nguon, S., Rathin, R., Sau, S., and Namazie, C. (2004). *The Macroeconomics of Poverty Reduction in Cambodia*. Asia-Pacific Regional Programme on the Macroeconomics of Poverty Reduction. Geneva: UNDP.

Berman, J. S. (1996). No place like home: anti-Vietnamese discrimination and nationality in Cambodia. *California Law Review*, 84(3), 817–74.

Better Factories Cambodia [BFC] (2016). *Garment Workers' Health and Nutrition Status, and Food Provision in Factories: A Study from Selected Enterprises in Cambodia*. Phnom Penh: BFC.

Beyenne, K. (2015). *Cambodia: Harnessing the Demographic Dividend—Lessons from Asian Tigers and Other Emerging Economies* UNDP DOI: 10.13140/RG.2.1.4771.0567

Blumtritt, A. (2013). Translocality and gender dynamics: The Pareja and the Thakhi system in Bolivia. *Bulletin of Latin American Research*, 32(1), 3–16.

Bonacich, P. (2007). Some unique properties of eigenvector centrality. *Social Networks*, 29(4), 555–64.

Bonacich, P., and Lloyd, P. (2001). Eigenvector-like measures of centrality for asymmetric relations. *Social Networks*, 23(3), 191–201.

Borras Jr, S. M., and Franco, J. C. (2013). Global land grabbing and political reactions 'from below'. *Third World Quarterly*, 34(9), 1723–47.

Bourke, L. (16 September, 2016) Inequality is the greatest threat to democracy, says President Barack Obama. *The Sunday Morning Herald*. Accessed 8/06/2018 at [http://www.smh.com.au]

Brady, H. E., Schlozman, K. L., and Verba, S. (2015). Political mobility and political reproduction from generation to generation. *The Annals of the American Academy of Political and Social Science*, 657(1), 149–73.

Brehm, W. C. (2016). The structures and agents enabling educational corruption in Cambodia. In Kitamura, Y., Edwards Jr, D. B., Williams, J. H., and Sitha, C. (eds), *The Political Economy of Schooling in Cambodia*. New York: Palgrave Macmillan, 99–119.

Brehm, W. C., and Silova, I. (2014). Hidden privatization of public education in Cambodia: Equity implications of private tutoring. *Journal for Educational Research Online*, 6(1), 94–116.

Bremner, J. and Hunter, L. (2014). Migration and the environment. *Population Bulletin* 69(1), 1–12.

Brickell, K. (2011a). Translocal geographies of 'home' in Siem Reap. In Brickell, K. and Datta, A. (eds), *Translocal Geographies: Spaces, Places, Connections*. London: Routledge.

Brickell, K. (2011b). The 'stubborn stain' on development: Gendered meanings of housework (non-) participation in Cambodia. *Journal of Development Studies*, 47(9), 1353–70.

Brickell, K. (2011c). 'We don't forget the old rice pot when we get the new one': Discourses on ideals and practices of women in contemporary Cambodia. *Signs: Journal of Women in Culture and Society*, 36(2), 437–62.

Brickell, K. (2012). Geopolitics of home. *Geography Compass*, 6(10), 575–88.

Brickell, K. (2014). 'The whole world is watching': intimate geopolitics of forced eviction and women's activism in Cambodia. *Annals of the Association of American Geographers*, 104(6), 1256–72.

Brickell, K., and Chant, S. (2010). 'The unbearable heaviness of being': reflections on female altruism in Cambodia, Philippines, The Gambia and Costa Rica. *Progress in Development Studies, 10*(2), 145–59.

Brickell, K., and Datta, A. (2011a). Introduction: translocal geographies. In Brickell, K. and Datta, A. (eds), *Translocal Geographies*. London: Routledge, 17–34.

Brickell, K. and Datta, A. (eds) (2011b). *Translocal Geographies*. London: Routledge.

Brinkley, J. (2013). Another stolen election in Cambodia. *World Affairs Journal*. 11 September 2013.

Brouwer, R., Akter, S., Brander, L., and Haque, E. (2007). Socioeconomic vulnerability and adaptation to environmental risk: a case study of climate change and flooding in Bangladesh. *Risk Analysis, 27*(2), 313–26.

Brown, J. C., and Purcell, M. (2005). There's nothing inherent about scale: Political ecology, the local trap, and the politics of development in the Brazilian Amazon. *Geoforum, 36*(5), 607–24.

Bryceson, D. F. (1996). Deagrarianization and rural employment in sub-Saharan Africa: A sectoral perspective. *World Development*, 24(1), 97–111.

Bulgaski, N. and Pred, D. (2010). *Formalizing Inequality: Land Titling in Cambodia*. Land Struggles: LRAN Briefing Paper Series.

Bylander, M. (2014). Borrowing across borders: Migration and microcredit in rural Cambodia. *Development and Change, 45*(2), 284–307.

Bylander, M. (2015a). Credit as coping: Rethinking microcredit in the Cambodian context. *Oxford Development Studies, 43*(4), 533–53.

Bylander, M. (2015b). Depending on the sky: Environmental distress, migration, and coping in rural Cambodia. *International Migration, 53*(5), 135–47.

Bylander, M., and Hamilton, E. R. (2015). Loans and leaving: migration and the expansion of microcredit in Cambodia. *Population Research and Policy Review, 34*(5), 687–708.

Cambodia Centre for Human Rights [CCHR] (2013). *'Workers' Rights Are Human Rights'. Policy Brief: The Garment Sector in Cambodia*. Phnom Penh: CCHR.

Campbell, C. (2019). Back from exile, Cambodia's opposition leader brings thousands onto the streets. *Phnom Penh Post*, 19/07/2019.

CARE International (2017). *'I Know I Cannot Quit.' The Prevalence and Productivity Cost of Sexual Harassment to the Cambodian Garment Industry*. Canberra: CARE Australia.

Casanova, E. M. D. (2013). Embodied inequality: The experience of domestic work in urban Ecuador. *Gender & Society, 27*(4), 561–85.

Castles, S., De Haas, H., and Miller, M. J. (2013). *The Age of Migration: International Population Movements in the Modern World*. London: Palgrave Macmillan.

Centre for Applied Non-Violent Actions and Strategies [CANVAS] (2016). *Analysis of the Situation in Cambodia*. Phnom Penh: CANVAS.

Centre for Ethnic and Racial Studies [CERS] working paper 2015. Accessed 17/02/2017 at [http://cers.leeds.ac.uk/working-papers].

Chandler, D. (2009). *A History of Cambodia*. London: Hachette UK.

Chhinh, N., and Millington, A. (2015). Drought monitoring for rice production in Cambodia. *Climate, 3*(4), 792–811.

Chim Charya, Srun Pithou, So Sovannarith, John McAndrew, Nguon Sokunthea, Pon Dorina, and Robin Biddulph (1998). *Learning from Rural Development Programmes in Cambodia—Working Paper 4* Phnom Pen: CDRI.

Chort, I., Gubert, F., and Senne, J. N. (2012). Migrant networks as a basis for social control: Remittance incentives among Senegalese in France and Italy. *Regional Science and Urban Economics, 42*(5), 858–74.

Clark, J. H. (2017). Feminist geopolitics and the Middle East: Refuge, belief, and peace. *Geography Compass, 11*(2), e12304. DOI 10.1111/gec3.12304

Cohen, S. A., and Gössling, S. (2015). A darker side of hypermobility. *Environment and Planning A: Economy and Space, 47*(8), 1666–79.

Conradson, D., and Mackay, D. (2007). Translocal subjectivities: mobility, connection, emotion. *Mobilities, 2*(2), 167–74.

Coulter, R., Van Ham, M., and Findlay, A. M. (2015). Re-thinking residential mobility linking lives through time and space. *Progress in Human Geography*, 0309132515575417.

Cresswell, T. (2010). Towards a politics of mobility. *Environment and planning D: Society and Space, 28*(1), 17–31.

Curran, S. R., and Saguy, A. C. (2013). Migration and cultural change: a role for gender and social networks? *Journal of International Women's Studies, 2*(3), 54–77.

Cuyvers, L., Reth, S. and Van de Bulke, D. (2009). The competitive position of a developing economy: The role of foreign direct investment in Cambodia, in Van de Bulke, D., Verbeke, A., and Yuan, W. (eds), *Handbook on Small Nations in the Global Economy: The Contribution of Multinational Enterprises to National Economic Success.* Cheltenham: Edward Elgar

Czymoniewicz-Klippel, M. T. (2013). Bad boys, big trouble subcultural formation and resistance in a Cambodian village. *Youth & Society, 45*(4), 480–99.

Daoud, N., O'Campo, P., Minh, A., Urquia, M. L., Dzakpasu, S., Heaman, M., Kaczorowski, J., Levitt, C., Smylie, J., and Chalmers, B. (2014). Patterns of social inequalities across pregnancy and birth outcomes: a comparison of individual and neighborhood socio-economic measures. *BMC Pregnancy and Childbirth, 14*(1), 393.

Davies, M. (2010). 'Cambodia: Country Case Study' in Leung, S., Bingham, B., and Davies, M. (eds), *Globalization and Development in the Mekong Economies.* Cheltenham: Edward Elgar Publishing.

Davis, K. F., Yu, K., Rulli, M. C., Pichdara, L., and D'Odorico, P. (2015). Accelerated deforestation driven by large-scale land acquisitions in Cambodia. *Nature Geoscience, 8*(10), 772.

Dawson, W. (2010). Private tutoring and mass schooling in East Asia: Reflections of inequality in Japan, South Korea, and Cambodia. *Asia Pacific Education Review, 11*(1), 14–24.

De Genova, N. P. (2002). Migrant 'illegality' and deportability in everyday life. *Annual Review of Anthropology, 31*(1), 419–47.

De Lisle, J., Seunarinesingh, K., Mohammed, R., and Lee-Piggott, R. (2017). Using an iterative mixed-methods research design to investigate schools facing exceptionally challenging circumstances within Trinidad and Tobago. *School Effectiveness and School Improvement, 28*(3), 406–42.

Dean, H. (1999). *Begging Questions: Street Level Economic Activity and Social Policy Failure.* Bristol: Policy Press.

Dercon, S. (2004). Growth and shocks: Evidence from rural Ethiopia. *Journal of Development Economics, 74*, 309–29.

Derks, A. (2008). *Khmer Women on the Move: Exploring Work and Life in Urban Cambodia.* Honolulu. University of Hawaii Press.

Deth, S. (2009). The geopolitics of Cambodia during the Cold War period *Explorations* 2009(9), 47–53.

Deth, S. U., and Bultmann, D. (2016). The afterglow of Hun Sen's Cambodia? Socioeconomic development, political change, and the persistence of inequalities. In Banpasirichote Wungaeo, C., Rehbein, B., and Wun'gaeo, S. (eds), *Globalization and Democracy in Southeast Asia.* London: Palgrave Macmillan, 87–109.

Devasahayam, T. W., Huang, S., and Yeoh, B. S. (2004). Southeast Asian migrant women: Navigating borders, negotiating scales. *Singapore Journal of Tropical Geography*, *25*(2), 135–40.

Dhaliwal, S., and Forkert, K. (2015). Deserving and undeserving migrants. *Soundings*, *61*(61), 49–61.

Diepart, J. C. (2015). *The Fragmentation of Land Tenure Systems in Cambodia: Peasants and the Formalization of Land Rights.* Working Paper Prepared for the Technical Committee on Land Tenure and Development.

Dixon, D. P. (2014). The way of the flesh: life, geopolitics and the weight of the future. *Gender, Place and Culture, 21*(2), 136–51.

Dorling, D. (2014). *Inequality and the 1.* London: Verso.

Dun, O. (2011). Migration and displacement triggered by floods in the Mekong Delta. *International Migration, 49*(s1), 200–23.

Durkheim, E. (1895 [2013]). *The Rules of the Sociological Method.* New York: Free Press.

Dwyer, M. B. (2015). The formalization fix? Land titling, land concessions and the politics of spatial transparency in Cambodia. *The Journal of Peasant Studies, 42*(5), 903–28.

Ebihara, M. (1968). *Svay, a Khmer Village in Cambodia* New York: Department of Columbia University Press.

Economist, the. (16 July 2014). Measuring inequality: A three-headed hydra. *The Economist.* Accessed 06/11/2018 at [https://www.economist.com/].

Edgar, A. (2002). *Cultural Theory: The Key Thinkers.* London: Routledge.

Edwards, P. (2007). *Cambodge: The Cultivation of a Nation, 1860–1945.* Honolulu: University of Hawaii Press.

Ek, G. (2013). Cambodia Environmental and Climate Change Policy Brief. SIDA's Helpdesk for Environmental and Climate Change. Accessed 04/04/2017 at [http://sidaenvironmenthelpdesk.se/]

Electoral Report Alliance [ERA] (2013). *Joint Report on the Conduct of the 2013 Cambodian Elections.* Phnom Penh: ERA.

Elmhirst, R. (2002). Daughters and displacement: migration dynamics in an Indonesian transmigration area. *Journal of Development Studies, 38*(5), 143–66.

Elmhirst, R. (2007). Tigers and gangsters: masculinities and feminised migration in Indonesia. *Population, Space and Place, 13*(3), 225–38.

Elmhirst, R. (2008). Gender and natural resource management: Livelihoods, mobility and interventions. In Resurreccion, B. P. and Elmhirst. R. (eds), *Multi-local Livelihoods, Natural Resource Management and Gender in Upland Indonesia.* London: Earthscan, 67–87.

Elmhirst, R. (2012). Methodological dilemmas in migration research in Asia: Research design, omissions and strategic erasures. *Area, 44*(3), 274–81.

Elmhirst, R., and Resurreccion, B. P. (2008). Gender, environment and natural resource management: New dimensions, new debates. In Resurreccion and Elmhirst (eds), *Gender and Natural Resource Management: Livelihoods, Mobility and Interventions*, London: Earthscan, 3–22.

Enfants & Developpement (2015). Reproductive and maternal health of garment workers in Kampong Speu. Accessed 02/01/2020 at https://docplayer.net/34077130-Reproductive-and-maternal-health-of-garment-workers-in-kampong-speu.html

Erskine, A., and Mackintosh, I. (1999). Why begging offends: Historical perspectives and continuities. In H. Dean (ed.), *Begging Questions: Street Level Economic Activity and Social Policy Failure* (pp. 27–42). Bristol: Policy Press.

Etchison, C. (2005). *After the Killing Fields; Lessons from the Cambodian Genocide.* Connecticut: Praeger.

Fabrega, H. (1971). Begging in a southeastern Mexican city. *Human Organization*, 30, 277–87.

Fafchamps, M. (1992). Solidarity networks in preindustrial societies: Rational peasants with a moral economy. *Economic Development and Cultural Change*, 41(1) 147–74.

Faist, T. (2016). Cross-border migration and social inequalities. *Annual Review of Sociology*, 42, 323–46.

Fein (1979). *Accounting for Genocide*. Chicago: University of Chicago Press

The Financial Times (9 April, 2014). Rising inequality is a blemish on Asia's growth story. Accessed 9 April 2014 at [http://www.ft.com]

Fischer, A. M. (2019). Bringing development back into development studies. *Development and Change*, 50(2), 426–44.

Fisher-Nguyen, K. (1994). Khmer proverbs: Images and rules. In Ebihara, M. M., Mortland, C. A., and Ledgerwood, J. (eds), *Cambodian Culture Since 1975: Homeland and Exile*. Ithaca: Cornell University Press.

Fleischer, M. (1995). *Beggars and Thieves: Lives of Urban Street Criminals*. Madison: University of Wisconsin Press.

Food and Agriculture Organisation [FAO] (2014). *Cambodia: Food and Agriculture Policy Decision Analysis*. Phnom Penh: FAO.

Fotopolous, T. (2016). Austrian elections, globalization, the massive rise of neo-nationalism and the bankruptcy of the left. *International Journal of Inclusive Democracy*, 12.

Franklin, B. (2007 [1754]). *Poor Richard's Almanac*. New York: Skyhorse Publishing.

Frewer, T. (2016). Cambodia's anti-Vietnam obsession. *The Diplomat*, 2 September 2016.

Friedman, M., and Friedman, R. (1980). *Free to Choose: A Personal Statement*. Boston: Houghton Mifflin Harcourt.

Fujii, T. (2013). Geographic decomposition of inequality in health and wealth: evidence from Cambodia. *The Journal of Economic Inequality*, 11(3), 373–92.

Future Forum (2016). *An Overview and Analysis of the Current Political Situation in Cambodia*. Accessed 27/02/2018 at [http://futureforum.asia].

Galtung, J. (1969). Violence, peace, and peace research. *Journal of Peace Research*, 6(3), 167–91.

Gartrell, A. (2010). 'A frog in a well': the exclusion of disabled people from work in Cambodia. *Disability & Society*, 25(3), 289–301.

Gartrell, A., and Hoban, E. (2013). Structural vulnerability, disability, and access to non-governmental organization services in rural Cambodia. *Journal of Social Work in Disability and Rehabilitation*, 12(3), 194–212.

Gartrell, A., and Hoban, E. (2016). 'Locked in space': Rurality and the politics of location. In Grech, S. and Soldatich, K. (eds), *Disability in the Global South*. Switzerland: Springer, Cham, 337–50.

Gerhartz, E. (2016). Mobility after war: re-negotiating belonging in Jaffa, Sri Lanka in Pellegrino, G. (ed.), *The Politics of Proximity: Mobility and Immobility in Practice*, 83–104.

German Agency for Technical Cooperation [GTZ] (2009). *Foreign Direct Investment (FDI) in Land in Cambodia*. Eschborn, Germany: GTZ.

Ghosh, J. (20 February, 2013). Inequality is the biggest threat to the world and needs to be tackled now. *The Guardian*. Accessed 18/06/2018 at [http://theguardian.com]

Glasser, I., and Bridgeman, R. (1999). *Braving the Street: The Anthropology of Homelessness*. New York, NY: Berghahn Books.

Glick Schiller, N., and Salazar, N. B. (2013). Regimes of mobility across the globe. *Journal of Ethnic and Migration Studies*, 39(2), 183–200.

Global Witness (2007). *Cambodia's Family Trees*. Washington DC: Global Witness.

Global Witness (2009). *Country for Sale: How Cambodia's Elite has Captured the Country's Extractive Industries*. London: Global Witness.

Global Witness (2010). *Shifting Sand: How Singapore's Demand for Cambodian Sand Threatens Ecosystems and Undermines Good Governance*. London: Global Witness.

Glover, T. D., and Hemingway, J. L. (2005). Locating leisure in the social capital literature. *Journal of leisure research*, 37(4), 387.

Goda, T., and García, A. T. (2017). The rising tide of absolute global income inequality during 1850–2010: is it driven by inequality within or between countries?. *Social Indicators Research*, 130(3), 1051–72.

Gottesman, E. (2003). *Cambodia after the Khmer Rouge: Inside the Politics of Nation Building*. New Haven: Yale University Press.

Grant, H. (2009). Manipulations of Cambodian nationalism: From French colonial rule to current polity. *Mapping Politics*, 1, 31–8.

Greiner, C., and Sakdapolrak, P. (2013). Translocality: Concepts, applications and emerging research perspectives. *Geography Compass*, 7(5), 373–84.

Hamano, T. (2010). Inequality and Disparity in Early Childhood Care and Education: The Case of Cambodia. *Proceedings* (9), 1–8.

Han, E. (2013). *Contestation and Adaptation: The Politics of National Identity in China*. Oxford: Oxford University Press.

Härkönen, J., Kaymakçalan, H., Mäki, P., and Taanila, A. (2012). Prenatal health, educational attainment, and intergenerational inequality: the Northern Finland Birth Cohort 1966 Study. *Demography*, 49(2), 525–52.

Harris, J. R., and Todaro, M. P. (1970). Migration, unemployment and development: a two-sector analysis. *The American Economic Review*, 60(1), 126–42.

Hayek, F. A. (2014 [1960]). *The Constitution of Liberty*. London: Routledge.

Heinonen, U. (2006). Environmental impact on migration in Cambodia: Water-related migration from the Tonle Sap Lake region. *International Journal of Water Resources Development*, 22(3), 449–62.

Helmers, K., and Jegillos, S. (2004). *Linkages between Flood and Drought Disasters and Cambodian Rural Livelihoods and Food Security: "How Can the CRC Community Disaster Preparedness Program Further Enhance Livelihood and Food Security of Cambodian Rural People in the Face of Natural Disasters?* Phnom Penh: Cambodian Red Cross.

Hill, H., and Menon, J. (2013). Cambodia: rapid growth with weak institutions. *Asian Economic Policy Review*, 8(1), 46–65.

Hill, P. S., and Ly, H. T. (2004). Women are silver, women are diamonds: conflicting images of women in the Cambodian print media. *Reproductive Health Matters*, 12(24), 104–15.

Hindle, S. (2004). Dependency, shame and belonging: Badging the deserving poor, c. 1550–1750. *Cultural and Social History*, 1, 6–35.

Hing, V., Lun, P., and Phann, D. (2011). *Irregular Migration from Cambodia: Characteristics, Challenges and Regulatory Approach*. Phnom Penh: CDRI.

Hing, V., Lun, P., and Phann, D. (2014). *The Impacts of Adult Migration on Children's Well-being: The Case of Cambodia*. Phnom Penh: Cambodia Development Resource Institute [CDRI].

Hokmeng, H., and Moolio, P. (2015). The impact of foreign aid on economic growth in Cambodia: A co-integration approach. *KASBIT Journal of Management and Social Science*, 8(1), 4–25.

Holsinger, D. B. and Jacob, W. J. (2008). *Inequality in Education: Comparative and International Perspectives*. Netherland: Springer.

Hong, R., and Them, R. (2015). Inequality in access to health care in Cambodia: socio-economically disadvantaged women giving birth at home assisted by unskilled birth attendants. *Asia Pacific Journal of Public Health, 27*(2), 1039–49.

Horner, R., and Hulme, D. (2019). From international to global development: New geographies of 21st century development. *Development and Change, 50*(2), 347–78.

Howe, L. (1998). Scrounger, worker, beggarman, cheat: the dynamics of unemployment and the politics of resistance in Belfast. *Journal of the Royal Anthropological Institute 4*(3), 531–50.

Hughes, C. (2001). Transforming oppositions in Cambodia. *Global Society, 15*(3), 295–318.

Hughes, C. (2007). Transnational networks, international organizations and political participation in Cambodia: Human rights, labour rights and common rights. *Democratization, 14*(5), 834–52.

Hughes, C. (2008). Cambodia in 2007: Development and dispossession. *Asian Survey, 48*(1), 69–74.

Hughes, C. (2015). Understanding the elections in Cambodia 2013. *AGLOS: Journal of Area-Based Global Studies*, Special Issue: Workshop and Symposium 2013-14, 1–20.

Hunsberger, C., Corbera, E., Borras Jr, S. M., de la Rosa, R., Eang, V., Franco, J. C., Herre, R., Sai S., Park, C., Pred, D., Sokheng, H., Spoor, M., Shwe, T., Kyaw, T., Ratha, T., Chayan, V., Woods, K., and Work, C. (2015). Land-based climate change mitigation, land grabbing and conflict: understanding intersections and linkages, exploring actions for change. *MOSAIC Working Paper Series No. 1.*

Hutt, D. (2016). The truth about Cambodia's anti-Vietnamese obsession. *The Diplomat.* 20 October 2016.

International Bank for Reconstruction and Development [IBRD] and World Bank. (2015). *Cambodian Agriculture in Transition: Opportunities and Risks Economic and Sector Work*. Report No.96308-kH. Washington, DC: World Bank Press.

International Campaign to Ban Landmines [ICBL]. (2014). Cambodia Country Report 2014. Retrieved from The Monitorwebsite:http://www.the-monitor.org

International Finance Corporation [IFC] (2015). *Cambodia Rice: Export Potential and Strategies*. Cambodia Agri-business Series, no.4. Phnom Penh: IFC.

International Fund for Agricultural Development [IFAD] (2014). *Investing in Rural People in Cambodia*. Rome: IFAD.

International Labour Organisation [ILO] (2012). *Action-oriented Research on Gender Equality and the Working and Living Conditions of Garment Factory Workers in Cambodia*. Geneva: ILO Press.

International Labour Organization [ILO] (2015). *Cambodia Garment Sector Bulletin, Issue 1*. Accessed 27/02/2018 at [http://ilo.org].

International Labour Organization [ILO] (2018). *Cambodia Garment and Footwear Sector Bulletin 1*. Issue 8. Phnom Penh: ILO.

Japan International Cooperation Agency [JICA] (2010). *Kingdom of Cambodia Study for Poverty Profiles in the Asian Region: Final Report*. Tokyo: JICA.

Jensen, A., and Richardson, T. (2007). New region, new story: Imagining mobile subjects in transnational space. *Space and Polity, 11*(2), 137–50.

Jensen, O. B. (2009). Flows of meaning, cultures of movements–urban mobility as meaningful everyday life practice. *Mobilities, 4*(1), 139–58.

Jimenez-Soto, E., Durham, J., & Hodge, A. (2014). Entrenched geographical and socio-economic disparities in child mortality: trends in absolute and relative inequalities in Cambodia. *PloS one, 9*(10), e109044. doi:10.1371/journal.pone.0109044

Jordan, B. (1999). Begging: The global context and international comparisons. In H. Dean (ed.), *Begging Questions: Street Level Economic Activity and Social Policy Failure*. Bristol: Policy Press, 43–62.

Justino, P. (2005). Empirical applications of multidimensional inequality analysis. *Poverty Research Unit at Sussex Working paper*. PRUS Working Paper No. 23. Accessed on 31/10/2018 at [http://citeseerx.ist.psu.edu/viewdoc/download?doi=10.1.1.362.3969&rep=rep1&type=pdf]

Justino, P., Litchfield, J., and Niimi, Y. (2004). Multidimensional Inequality: An Empirical Application to Brazil. Paper prepared for the 28th General Conference of The International Association for Research in Income and Wealth Cork, Ireland, 22–28 August 2004.

Kabeer, N. (1997). Women, wages and intra-household power relations in urban Bangladesh. *Development and Change, 28*(2), 261–302.

Kalab, M. (1968). Study of a Cambodian village. *The Geographical Journal*, 521–37.

Kalir, B. (2013). Moving subjects, stagnant paradigms: can the 'mobilities paradigm' transcend methodological nationalism?.*Journal of Ethnic and Migration Studies, 39*(2), 311–27.

Kashyap, A. (2015). *'Work Faster or Get Out': Labor Rights Abuses in Cambodia's Garment Industry*. Phnom Penh: Human Rights Watch.

Kassah, A. K. (2008). Begging as work: A study of people with mobility difficulties in Accra, Ghana. Disability and Society, 23, 163–70.

Keo, S. (2015). The impact of climate change on agricultural production in north-west Cambodia. Presented at the *Conference on International Research on Food Security, Tropentag 2015*, September 16–18, Berlin, Germany

Kheam, T. and Treleaven, E. (2013). *Women and Migration in Cambodia: A Further Analysis of the Cambodian Rural–Urban Migration Project (CRUMP)* Phnom Penh, Cambodia: UNFPA and National Institute of Statistics.

Khy, S. (2017). Deportations of Vietnamese dropped last year. *The Cambodia Daily*. 23 January 2017.

King, R., and Skeldon, R. (2010). 'Mind the gap!' Integrating approaches to internal and international migration. *Journal of Ethnic and Migration Studies, 36*(10), 1619–46.

Kolben, K. (2004). Trade, monitoring, and the ILO: Working to improve conditions in Cambodia's garment factories. *Yale Human Rights and Development*. LJ(7), 79–109.

Krishna, A. (2010). *One Illness Away: Why People Become Poor and How They Escape Poverty*. New York, NY: Oxford University Press.

Kung, P. (2012). How has Cambodia achieved political reconciliation? In Sothirak, P., Wade, G. and Hong, M. (eds), *Cambodia: Progress and Challenges since 1991*. Singapore: Institute of Southeast Asian Studies.

Kurien, J. (2007). Shocking reality: Cambodia's electro-fishing. *Samudra Report* (28) Phnom Penh: ICSF.

Kuznets, S. (1955). Economic growth and income inequality. *American Economic Review* 1, 1–28.

Law, R. (1999). Beyond 'women and transport': towards new geographies of gender and daily mobility. *Progress in Human Geography, 23*(4), 567–88.

Lawrance, B. N. (2007). *Locality, Mobility, and' Nation': Periurban Colonialism in Togo's Eweland, 1900–1960* (Vol. 31). New York: University of Rochester Press.

Lawreniuk, S. (2016). Rural–urban linkages in the Cambodian migration system. In Brickell, K. and Springer, S. (eds), *The Handbook of Contemporary Cambodia*. London: Routledge, 202–11.

Lawreniuk, S. (2017). The ties that bind: Rural–urban linkages in the Cambodian migration system. In Brickell, K. and Springer, S. (eds), *The Handbook of Contemporary Cambodia*. London: Routledge, pp. 202–11.

Lawreniuk, S. (2019). Intensifying political geographies of authoritarianism: Garment worker struggles in neoliberal Cambodia. *Annals of the Association of American Geographers* (forthcoming). DOI: 10.1080/24694452.2019.1670040.

Lawreniuk, S., and Parsons, L. (2017). Mother, grandmother, migrant: Elder translocality and the renegotiation of household roles in Cambodia. *Environment and Planning A*, *49*(7), 1664–83.

Lawreniuk, S. and Parsons, L. (2018). For a few dollars more: Towards translocal mobilities of labour activism in Cambodia. *Geoforum*. *92*, 26–35.

Layton, R., and MacPhail, F. (2013). *Gender Equality in the Labour Market in Cambodia*. Manila: Asian Development Bank.

Le Billon, P. (2000). The political ecology of transition in Cambodia 1989–1999: War, peace and forest exploitation. *Development and Change*, *31*(4), 785–805.

Le Billon, P., and Springer, S. (2007). Between war and peace: Violence and accommodation in the Cambodian logging sector. In De Jong, W., Donovan, D., and Abe, K. (eds). *Extreme Conflict and Tropical Forests*. Rotterdam: Springer Netherlands, 17–36.

Leng Heng An (2014). Report Prepared for the Asian Disaster Reduction Centre, February 2014. Accessed on 27/01/2020 at [https://www.adrc.asia]

Levin, H. M. (1976). Educational opportunity and social inequality in Western Europe. *Social Problems*, *24*(2), 148–72.

Levitt, P., and Lamba-Nieves, D. (2011). Social remittances revisited. *Journal of Ethnic and Migration Studies*, *37*(1), 1–22.

Lewis, A. (2015). *Historical Patterns of the Racialisation of Vietnamese in Cambodia, and Their Relevance Today*.

Lewis, W. A. (1954). Economic development with unlimited supplies of labour. *The Manchester School*, *22*(2), 139–91.

Li, Y., and Wei, Y. D. (2014). Multidimensional inequalities in health care distribution in provincial China: A case study of Henan Province. *Tijdschrift voor economische en sociale geografie*, *105*(1), 91–106.

LICADHO (2009) *Land Grabbing and Poverty in Cambodia: the Myth of Development*. Phnom Penh: LICADHO.

Liese, B., Isvilalonda, S., Nguyen Tri, K., Nguyen Ngoc, L., Pananurak, P., Pech, R. Shwe, T.N., Sombounkhanh, Möllman, T., and Zimmer, Y. (2014). *Economics of Southeast Asian Rice Production* Agri-Benchmark Working Paper 2014/1 accessed at http://www.agribenchmark.org/fileadmin/Dateiablage/B-Cash-Crop/Reports/Report-2014-1-rice-FAO.pdf

Lim, S. (2007). *Youth Migration and Urbanisation in Cambodia*. Phnom Penh: Cambodia Development Resource Institute.

Lindley, A. (2009). The early-morning phonecall: Remittances from a refugee diaspora perspective. *Journal of Ethnic and Migration Studies*, *35*(8), 1315–34.

Lindley, A. (2010). *The Early Morning Phone Call: Somali Refugees' Remittances* (Vol. 28). Oxford: Berghahn Books.

Lipes, J. (2013). Striking garment workers join CNRP protest. *Radio Free Asia*. 26 December 2013.

Louth, J. (2015). Neoliberalising Cambodia: The production of capacity in Southeast Asia. *Globalizations*, *12*(3), 400–19.

Macdonald, M. (1958). *Angkor, with One Hundred and Twelve Photographs.* London: Jonathan Cape.

Mackay, A., Mussida, C. and Veruete, L. (2016). *The Nature of Youth Employment in Cambodia: Informal Activity Continues to Dominate Despite Consistent Economic Growth.* Draft paper made available by the Associazione Italiana Economisti Lavoro. Accessed 06/11/2018 at [http://www.aiel.it/page/publications.php]

Madhur, S. (2013). *Sustaining Cambodia's Development Miracle: What's Next?* Paper presented at Cambodia Outlook Conference, 20 February 2013, Phnom Penh.

Marcus, G. E. (1995). Ethnography in/of the world system: The emergence of multi-sited ethnography. *Annual Review of Anthropology,* 24(1), 95–117.

Marger, M. (1999). *Social Inequality: Patterns and Processes.* Mountain View, California: Mayfield Publishing Company.

Marks, D., Sirithet, A., Rakyuttitham, A., Wulandari, S., Chomchan, S., and Samranjit, P. (2015). *Land Grabbing and Impacts to Small Scale Farmers in Southeast Asia Sub-Region.* Bangkok: Local Act Thailand.

Marschke, M., Lykhim, O., and Kim, N. (2014). Can local institutions help sustain livelihoods in an era of fish declines and persistent environmental change? A Cambodian case study. *Sustainability,* 6(5), 2490–2505.

Massey, D. S. (1990). Social structure, household strategies, and the cumulative causation of migration. *Population Index* 56(1) 3–26.

Massey, D., Rafique, A., and Seeley, J. (2010). Begging in rural India and Bangladesh. *Economic and Political Weekly,* 45, 64–71.

Mauss, M. (2002 [1925]). *The Gift: The Form and Reasons for Exchange in Archaic Societies.* London: Routledge.

May, T. (2013). At the border, Rainsy plays old tune. *The Phnom Penh Post.* 25 July 2013.

McKay, D. (2007). 'Sending dollars shows feeling'—emotions and economies in Filipino migration. *Mobilities,* 2(2), 175–94.

McKeown, A. (2004). Global Migration 1846–1940. *Journal of World History,* 15(2), 155–89.

Meas, N. (1999). *Towards Restoring Life: Cambodian Villages.* Phnom Penh: JSRC Printing House.

Meyfroidt, P., Lambin, E. F., Erb, K. H., and Hertel, T. W. (2013). Globalization of land use: Distant drivers of land change and geographic displacement of land use. *Current Opinion in Environmental Sustainability,* 5(5), 438–44.

Milanovic, B. (2016). *Global Inequality: A New Approach for the Age of Globalization.* Cambridge: Harvard University Press.

Millar, P. (2016). Race to the bottom: How Cambodia's opposition is targeting ethnic Vietnamese. *Southeast Asia Globe,* 21 October 2016.

Milne, S. (2012). Grounding forest carbon: Property relations and avoided deforestation in Cambodia. *Human Ecology,* 40(5), 693–706.

Milne, S., and Adams, B. (2012). Market masquerades: uncovering the politics of community-level payments for environmental services in Cambodia. *Development and Change,* 43(1), 133–58.

Milne, S. (2013). Under the leopard's skin: Land commodification and the dilemmas of indigenous communal title in upland Cambodia. *Asia Pacific Viewpoint,* 54(3), 323–39.

Milne, S. (2015). Cambodia's unofficial regime of extraction: Illicit logging in the shadow of transnational governance and investment. *Critical Asian Studies,* 47(2), 200–28.

Milne, S., and Mahanty, S. (eds) (2015). *Conservation and Development in Cambodia: Exploring Frontiers of Change in Nature, State and Society.* London: Routledge.

Ministry of Environment [MoE] and British Broadcasting Corporation [BBC]. (2011). *Understanding Public Perceptions of Climate Change in Cambodia Ministry of Environment.* Phnom Penh: Climate Change Department.

National Institute of Statistics [NIS]. (2010). *Labour and Social Trends in Cambodia.* Phnom Penh: NIS.

Neef, A., Touch, S., and Chiengthong, J. (2013). The politics and ethics of land concessions in rural Cambodia. *Journal of Agricultural and Environmental Ethics, 26*(6), 1085–1103.

Neelsen, J. P. (1975). Education and social mobility. *Comparative Education Review, 19*(1), 129–43.

Newman, M. E. (2008). The mathematics of networks, in *The New Palgrave Encyclopedia of Economics.* London: Palgrave

Newton, L. (2005). It is not a question of being anti-immigration: Categories of deserved-ness in immigration policy making. In Schneider, A. and Ingram, H. (eds), New York: State University of New York Press. *Deserving and Entitled: Social Constructions and Public Policy*, 139–67.

Ninh, T. H. T. (2017). *Race, Gender and Religion in the Vietnamese Diaspora.* London: Palgrave.

Nuon, V., and Serrano, M. R. (2010). *Building Unions in Cambodia: History, Challenges, Strategies.* Singapore: Freidrich Ebert Stiftung.

Oakes, T., and Schein, L. (eds) (2006). *Translocal China: Linkages, identities and the Reimagining of Space.* London: Routledge.

Oeur, I. L., Sopha, A., and McAndrew, J. (2012). Understanding social capital in response to flood and drought: A study of five villages in two ecological zones in Kampong Thom Province, in Pellini, A. (ed.), *Engaging for the Environment.* Phnom Penh: The Learning Institute, 60–84.

Ogawa, Y. (2004). Are agricultural extension programs gender sensitive? Cases from Cambodia. *Gender, Technology and Development, 8*(3), 359–80.

Oldenburg, C., and Neef, A. (2014). Reversing land grabs or aggravating tenure insecurity? Competing perspectives on economic land concessions and land titling in Cambodia. *Law and Development Review, 7*(1), 49–77.

Olvera, L. D., Mignot, D., and Paulo, C. (2004). Daily mobility and inequality: the situation of the poor. *Built Environment, 30*(2), 153–60.

Ovesen, J., and Trankell, I. B. (2014). Symbiosis of microcredit and private moneylending in Cambodia. *The Asia Pacific Journal of Anthropology, 15*(2), 178–96.

Oxfam (2016). *An Economy for the 1%: How Privilege and Power in the Economy Drive Extreme Inequality and How This Can be Stopped.* Oxford Briefing Paper 210. Accessed 18/06/2018 at [http://www.oxfam.org]

Pain, R. (2014). Everyday terrorism: Connecting domestic violence and global terrorism. *Progress in Human Geography, 38*(4), 531–50.

Palloni, A., Massey, D. S., Ceballos, M., Espinosa, K., and Spittel, M. (2001). Social capital and international migration: A test using information on family networks. *American Journal of Sociology, 106*(5), 1262–98.

Parnell, T. (2015). Story-telling and social change: A case study of the Prey Lang community network, in Milne, S. and Mahanty, S. (eds), *Conservation and Development in Cambodia: Exploring New Frontiers of Change in Nature, State and Society.* London: Routledge, 258–79.

Parsons, L. (2016). Mobile inequality: Remittances and social network centrality in Cambodian migrant livelihoods. *Migration Studies, 4*(2), 154–81.

Parsons, L. (2017a). Environmental risk and contemporary resilience strategies in rural Cambodia. In Brickell, K. and Springer, S. (eds), *The Handbook of Contemporary Cambodia*, 146–56.

Parsons, L. (2017b). Multi-scalar inequality: Structured mobility and the narrative construction of scale in translocal Cambodia. *Geoforum, 85,* 187–96.

Parsons, L. and Lawreniuk, S. (2016). The village of the damned? myths and realities of structured begging behaviour in and around Phnom Penh *The Journal of Development Studies, 52*(1), 36–52

Parsons, L. and Lawreniuk, S. (2017a). 'A viscous cycle: low motility amongst Phnom Penh's highly mobile cyclo riders' *Mobilities, 12*(5), 646–62.

Parsons, L., and Lawreniuk, S. (2017b). Love in the time of Nokia: Cultural change as compromise in a Cambodian migrant enclave. *Population, Space and Place, 23*(3), 1–18.

Parsons, L., and Lawreniuk, S. (2018). Seeing like the stateless: Documentation and the mobilities of liminal citizenship in Cambodia. *Political Geography, 62,* 1–11.

Parsons, L., Lawreniuk, S., and Pilgrim, J. (2014). Wheels within wheels: Poverty, power and patronage in the Cambodian migration system. *The Journal of Development Studies, 50*(10), 1362–79.

Partnering to Save Lives [PSL] (2014). *Baseline Survey Report: Reproductive, Maternal and Neonatal Health Knowledge.* Phnom Penh: PSL.

Pershai, A. (2008). Localness and mobility in Belarusian nationalism: The tactic of Tuteishaść. *Nationalities Papers, 36*(1), 85–103.

Peters, M. A. (2018). The end of neoliberal globalisation and the rise of authoritarian populism. *Educational Philosophy and Theory 50*(4), 323–5.

Phong, K. and Solá, J. (2015) *Mobile Phones and Internet in Cambodia.* Phnom Penh: The Open Institute.

Pickett, R. and Wilkinson, K. (2009). *The Spirit Level: Why More Equal Societies Almost Always Do Better.* New York: Bloomsbury.

Piguet, E. (2010). Linking climate change, environmental degradation, and migration: A methodological overview. *Wiley Interdisciplinary Reviews: Climate Change, 1*(4), 517–24.

Piketty, T. (2014). *Capital in the 21st Century.* Cambridge: Harvard University Press.

Platt, M., Yen, K. C., Yeoh, B. S., Luymes, G., and Lam, T. (2016). *Translocal Subjectivities within Households 'in Flux' in Indonesia.* Working Paper, 46, Migrating out of Poverty Research Programme Consortium. Accessed on 07/11/2018 at [http://migratingoutofpoverty.dfid.gov.uk/]

Platteau, J-P. (2000). *Institutions, Social Norms and Economic Development.* The Netherlands: Harwood.

Platteau, J-P., and Abraham, A. (2002). Participatory development in the presence of endogenous community imperfections. *Journal of Development Studies, 39*(2), 104–36.

Porst, L., and Sakdapolrak, P. (2017). How scale matters in translocality: Uses and potentials of scale in translocal research. *Erdkunde, 71*(2), 111–26.

Potts, D. (2010). *Circular Migration in Zimbabwe and Contemporary sub-Saharan Africa.* Rochester, NY: Boydell and Brewer.

Powell, J. (12 February 2014). '70 Great Quotations about the Glory of Honest Work and Achievement'. *Forbes.* Accessed on 18/06/2018 at [http://www.forbes.com]

Putnam, R. D. (1995). Bowling alone: America's declining social capital. *Journal of Democracy, 6*(1), 65–78.

Rahut, D. B., and Micevska Scharf, M. (2012). Non-farm employment and incomes in rural Cambodia. *Asian-Pacific Economic Literature, 26*(2), 54–71.

Ramanthan, U. (2008). Ostensible poverty, beggary and the law. *Economic and Political Weekly, 33*, 33–44.

Resurreccion, B. P. (2005). Women in-between gender, transnational and rural–urban mobility in the Mekong region. *Gender, Technology and Development, 9*(1), 31–56.

Resurreccion, B. P., Sajor, E. E., and Sophea, H. (2008). *Gender Dimensions of the Adoption of the System of Rice Intensification (SRI) in Cambodia.* Phnom Penh: Oxfam America.

Rigg, J. (2012). *Unplanned Development: Tracking Change in South-East Asia.* London: Zed Books.

Rigg, J. (2013). From rural to urban: A geography of boundary crossing in Southeast Asia. *TRaNS: Trans-Regional and-National Studies of Southeast Asia, 1*(1), 5–26.

Rigg, J., Nguyen, T. A., and Luong, T. T. H. (2014). The texture of livelihoods: Migration and making a living in Hanoi. *Journal of Development Studies, 50*(3), 368–82.

Rindfuss, R. R., Piotrowski, M., Entwisle, B., Edmeades, J., and Faust, K. (2012). Migrant remittances and the web of family obligations: Ongoing support among spatially extended kin in North-east Thailand, 1984–94. *Population Studies, 66*(1), 87–104.

Ringmar, E. (2016). *The Nation-State As Failure: Nationalism and Mobility, in India and Elsewhere.* Lund University Working Paper Series 2016: 5.

Rogaly, B., and Thieme, S. (2012). Experiencing space–time: The stretched lifeworlds of migrant workers in India. *Environment and Planning A, 44*(9), 2086–2100.

Ros, C. (2016). Public school salaries set to increase. *The Khmer Times,* 2 January. Accessed 01/11/2018 at [https://www.khmertimeskh.com/news/19814/public-school-salaries-set-to-increase]

Ros, C. (2017). More than 3500 foreign nationals deported in 2016. *The Khmer Times,* 4 May 2017.

Ros, B., Chhun, C., and Phirun, N. (2011). *Agricultural Development and Climate Change: The Case of Cambodia.* Phnom Penh: CDRI.

Rumsby, C. (2015). *Acts of Citizenship and Alternative Perspectives on Voice among Stateless Vietnamese Children in Cambodia* Institute of Statelessness and Inclusion Working Paper Series No. 2015/04 accessed 17/02/2017 at [http://www.institutesi.org].

Schaffer, D. W. (2017). *Quantitative Ethnography.* Madison: Cathcart Press.

Schneider, A. E. (2011). What shall we do without our land? Land grabs and resistance in rural Cambodia. Paper presented at the *International Conference on Global Land Grabbing,* 6–8 April 2011.

Schneider, J., and Crul, M. (2010). New insights into assimilation and integration theory: Introduction to the special issue. *Ethnic and Racial Studies, 33*(7), 1143–8.

Schulze, H. (1998). *States, Nations and Nationalism: From the Middle Ages to the Present.* Wiley-Blackwell.

Scott, J. C. (1972). Patron–client politics and political change in Southeast Asia. *The American Political Science Review, 66*(1), 91–113.

Scurrah, N., and Hirsch, P. (2015). *The Political Economy of Land Governance in Cambodia.* Phnom Penh: Mekong Region Land Governance.

Seiff, A. (2013). Hero's welcome for returning Rainsy. *The Phnom Penh Post,* 19 July 2013.

Sentamu, J., Archbishop of York. (31 October 2015). 'We must use the living wage to slay the ogre of income inequality'. *The Guardian.* Accessed 06/11/2018 at [http://theguardian.com]

Sheng, Y. (2012). The challenges of promoting productive, inclusive and sustainable urbanization. In Sheng, Y. K. and Thuzari, M. (eds), *Urbanization in Southeast Asia Issues and Impact.* Singapore: Institute of Southeast Asian Studies.

Sherchan, D. (2015). *The Bitter Taste of Sugar: Displacement and Dispossession in Oddar Meanchey Province.* Phnom Penh: Action Aid Cambodia and Oxfam.

Shleifer, A. and Vishny, R.W. (1998). *The Grabbing Hand: Government Pathologies and their Cures.* Cambridge: Harvard University Press.

Shukla, N. (2013). South Asian migration to the United States: diasporic and national formations. In Chatterji, J. and Washbrook, D. (eds), *The Routledge Handbook of the South Asian Diaspora.* London: Routledge.

Siciliano, G., Urban, F., Tan-Mullins, M., Pichdara, L., and Kim, S. (2016). The political ecology of Chinese large dams in Cambodia: Implications, challenges and lessons learnt from the Kamchay Dam. *Water, 8*(9), 405.

Silvey, R. (2001). Migration under crisis: Disaggregating the burdens of household safety nets. *Geoforum, 32*(1), 33–45.

Sim, S. (2004). *Report on the Health Status of Women Workers in the Cambodian Garment Industry.* Phnom Penh: Womyn's Agenda for Change.

Singh, S. (2014). Borderland practices and narratives: Illegal cross-border logging in north-eastern Cambodia. *Ethnography, 15*(2), 135–59.

Skeldon, R. (1997). *Migration and Development: a Global Perspective.* London: Routledge.

Smith, S. (2012). Intimate geopolitics: Religion, marriage, and reproductive bodies in Leh, Ladakh. *Annals of the Association of American Geographers, 102*(6), 1511–28.

Sophal, C. (2008). Impact of high food prices in Cambodia. *CDRI Policy Brief, 2,* 1–6.

Sothearith T, and Sovannarith S. (2009). *Impact of Hiked Prices of Food and Basic Commodities on Poverty in Cambodia: Empirical Evidence from CBMS from Five Villages.* Cambodia: CMBS. *Monitoring the Impacts of Economic Crises Using CBMS.* Proceedings of the 2008 CBMS Network Meeting, 9–12 December 2008, pp. 217–48.

Springer, S. (2010). Neoliberal discursive formations: on the contours of subjectivation, good governance, and symbolic violence in post-transitional Cambodia. *Environment and Planning D: Society and Space, 28*(5), 931–50.

Springer, S. (2011). Articulated neoliberalism: the specificity of patronage, kleptocracy, and violence in Cambodia's neoliberalization. *Environment and Planning A, 43*(11), 2554–70.

Stark, O., and Bloom, D. E. (1985). The new economics of labour migration. *The American Economic Review,* 173–8.

Stirbu, M., Polack, E., and Bun P. (2010). Climate change, children and youth in Cambodia: Successes, challenges and policy implications. *IDS Policy Brief August 2010.* Accessed 06/11/2018 at [http://www.childreninachangingclimate.org].

Swanson, K. (2010). *Begging as a Path to Progress: Indigenous Women and Children and the Struggle for Ecuador's Urban Spaces (Vol. 2).* Athens: University of Georgia Press.

Swyngedouw, E. (2004). Scaled geographies: Nature, place, and the politics of scale. *Scale and Geographic Inquiry: Nature, Society, and Method,* London: Blackwell, 129–153.

Taing, V. (2015). Nationalism: US, Cambodia voters decide. *Khmer Times,* 31 October 2016.

Tarrant, A. (2010). Constructing a social geography of grandparenthood: A new focus for intergenerationality. *Area, 42*(2), 190–7.

Taylor, E. J. (1999). The new economics of labour migration and the role of remittances in the migration process. *International Migration, 37*(1), 63–88.

Teehan, S., Mom Kunthear, and Taking Vida (2014). A brand new strategy. *Phnom Penh Post,* 18/09/2014.

ter Horst, G. C. H. (2008). *Weaving into Cambodia: Trade and Identity Politics in the (post)-Colonial Cambodian Silk Weaving Industry.* Doctoral dissertation presented to Vrije University, Amsterdam.

Tesfahuney, M. (1998). Mobility, racism and geopolitics. *Political Geography*, *17*(5), 499–515.

Thomas, P. (2005). *Poverty Reduction and Development in Cambodia: Enabling Disabled People to Play a Role*. London: UK Department for International Development.

Todaro, M. P. (1969). A model of labor migration and urban unemployment in less developed countries. *The American Economic Review*, *59*(1), 138–48.

Tong, K. and Sry, B. (2011). *Poverty and Environment Links: The Case of Rural Cambodia*. Phnom Penh: CDRI

Toyota, M., Yeoh, B. S., and Nguyen, L. (2007). Bringing the 'left behind' back into view in Asia: a framework for understanding the 'migration–left behind nexus'. *Population, Space and Place*, *13*(3), 157–61.

Truong, T. D., and Gasper, D. (2008). Trans-local livelihoods and connections: Embedding a gender perspective into migration studies. *Gender, Technology and Development*, *12*(3), 285–302.

Tsuda, T. (1998). The stigma of ethnic difference: The structure of prejudice and ' discrimination' toward Japan's new immigrant minority. *Journal of Japanese Studies*, *24*(2), 317–59.

Un, K., and So, S. (2009). Politics of natural resource use in Cambodia. *Asian Affairs: An American Review*, *36*(3), 123–38.

United Nations Capital Development Fund [UNCDF] (2017). *Remittances as a Driver of Women's Financial Inclusion*. Asia Pacific Regional Office: UNCDF and AusAid.

United Nations Development Program [UNDP] (2013). *Industry-Agriculture Linkages: Implications for Rice Policy*. Discussion paper no. 9, 2013. Phnom Penh: UNDP.

United Nations Development Program [UNDP] (2016). Human Development Report 2016: Cambodia. 2018 Statistical Update. Accessed on 31/10/2018 at [http://hdr.undp.org/sites/all/themes/hdr_theme/country-notes/KHM.pdf]

United Nations Development Program [UNDP] (2018). UN Human Development Report 2018: Cambodia. Accessed on 14/01/2020 at [http://hdr.undp.org]

United Nations International Children's Emergency Fund [UNICEF] (2017). *Executive Summary Study on the Impact of Migration on Children in the Capital and Target Provinces, Cambodia*. Phnom Penh: UNICEF.

United States Agency for International Development [USAID] (2011). *Property Rights and Resource Governance: Cambodia*. Washington: USAID.

United States Agency for International Development [USAID] (2013). *School Dropout Prevention Pilot (SDPP) Program. Summary Annual Progress Report, October 1, 2012— September 30, 2013*. Washington: World Bank Press.

United States State Department (2015). *Cambodia Investment Climate Statement*. Accessed on 25/10/2017 at [http//www.state.gov]

Unteroberdoerster, O. (2014). *Cambodia: Entering a New Phase of Growth*. Washington: International Monetary Fund.

Van den Brink, M., and Benschop, Y. (2012). Slaying the seven-headed dragon: The quest for gender change in academia. *Gender, Work and Organization*, *19*(1), 71–92.

Vannarith, C. (2015). *How Cambodian Nationalism is Driving Border Disputes with Vietnam*. East Asia Forum. Accessed 27/02/2018 at [http://www.eastasiaforum.org/].

Varis, O. (2008). Poverty, economic growth, deprivation, and water: the cases of Cambodia and Vietnam. *AMBIO: A Journal of the Human Environment*, *37*(3), 225–31.

Vigil, S. (2016). *Without Rain or Land, Where Will Our People Go? Climate Change, Land Grabbing, and Human Mobility: Insights from Senegal and Cambodia*. Colloquium Paper No. 60. The Hague: International Institute of Social Studies [ISS].

Volpicelli, S. (2015). *Who's Afraid of...Migration?: A New European Narrative of Migration*. Istituto Affari Internazionali [IAI] working paper series *15*(32), 1–26.

Warner, K., Hamza, M., Oliver-Smith, A., Renaud, F., and Julca, A. (2010). Climate change, environmental degradation and migration. *Natural Hazards, 55*(3), 689–715.

Weiss, H. (2007). *Begging and Alms Giving in Ghana: Muslim Positions towards Poverty and Distress*. Research Report 133. Uppsala: Nurdiska Afrikainstitutet.

Wilkinson, R., and Pickett, K. (2009). *The Spirit Level: Why Greater Equality Makes Societies Stronger*. New York: Bloomsbury Publishing USA.

Winsemius, H. C., Jongman, B., Veldkamp, T., Hallegatte, S., Bangalore, M., and Ward, P. (2015). *Disaster Risk, Climate Change, and Poverty: Assessing the Global Exposure of Poor People to Floods and Droughts*. Police Research Working Paper 7480. Washington: World Bank Group.

World Bank (2014). *Where Have All the Poor Gone: Cambodia Poverty Assessment 2013*. Washington: World Bank Press.

World Bank (2016). *Leveraging the Rice Value Chain for Poverty Reduction: In Cambodia Laos PDR and Myanmar*. Economic and Sector Work Report No. 105285-EAP. Washington: World Bank.

World Bank (2018a). Cambodia Country Data. Accessed on 31/10/2018 at [https://data.worldbank.org/country/Cambodia]

World Bank (2018b). *Cambodia Economic Update, April 2018: Recent Economic Developments and Outlook*. Phnom Penh: World Bank.

World Bank Group (2017). *Global Economic Prospects: A Fragile Recovery*. Washington: IBRD and World Bank.

The WorldFish Centre (2009). *Climate Change and Fisheries: Vulnerability and Adaptation in Cambodia*. Penang: WorldFish.

World Health Organisation [WHO] (2015). *State of Inequality: Reproductive, Maternal, Newborn and Child Health*. Luxembourg: World Health Organisation.

Xiang, B. (2013). Multi-scalar ethnography: An approach for critical engagement with migration and social change. *Ethnography, 14*(3), 282–99.

Yagura, K. (2005). Imperfect markets and emerging landholding inequality in Cambodia. *The Japanese Journal of Rural Economics, 7*, 30–48.

Yagura, K. (2015). Intergenerational land transfer in rural Cambodia since the late 1980s: Special attention to the effect of labor migration. *Southeast Asian Studies, 4*(1), 3–42.

Yeoh, B. S., and Huang, S. (2014). Singapore's changing demography, the eldercare predicament and transnational 'care' migration. *TRaNS: Trans-Regional and-National Studies of Southeast Asia, 2*(2), 247–69.

Yeoh, B. S., Graham, E., and Boyle, P. J. (2002). Migrations and family relations in the Asia Pacific region. *Asian and Pacific Migration Journal, 11*(1), 1–11.

Yeoh, S. A. B., and Huang, S. L. S. (2010). *Mothers on the Move: Children's Education and Transnational Mobility in Global-City*. Singapore. Routledge.

Yu, B., and Fan, S. (2011). Rice production response in Cambodia. *Agricultural Economics, 42*(3), 437–50.

Yu, P., and Berryman, D. L. (1996). The relationship among self-esteem, acculturation, and recreation participation of recently arrived Chinese immigrant adolescents. *Journal of Leisure Research, 28*(4), 251.

Index

Note: Tables and figures are indicated by an italic '*t*' and '*f*', respectively, following the page number.

For the benefit of digital users, indexed terms span two pages (e.g., 52–53) may, on occasion, appear on only one of those pages.